CONSTRUCTION TECHNOLOGY

£25.99

CONSTRUCTION TECHNOLOGY

THIRD EDITION

ROY CHUDLEY *MCIOB*

Revised by
ROGER GREENO *BA (HONS), FCIOB*

PEARSON
Longman

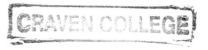

Harlow, England • London • New York • Boston • San Francisco • Toronto • Sydney • Singapore • Hong Kong
Tokyo • Seoul • Taipei • New Delhi • Cape Town • Madrid • Mexico City • Amsterdam • Munich • Paris • Milan

Pearson Education Limited
Edinburgh Gate
Harlow
Essex CM20 2JE
England

and Associated Companies throughout the world

Visit us on the World Wide Web at:
www.pearsoned.co.uk

First published (as *Construction Technology*) 1973 (Volume 1), 1974 (Volume 2)
Second edition 1987
Third edition (published as a single volume, with revisions by Roger Greeno) 1999
Reprinted 2002 (revised edition), 2003 (twice) (third edition update), 2004 (twice)

British Library Cataloguing in Publication Data
A catalogue entry for this title is available from the British Library

ISBN 0-582-31616-2

15   14   13   12   11   10   9
08   07   06   05   04

Set by 35 in Ehrhardt 10/12 and Futura
Printed in Malaysia, VP

CONTENTS

PREFACES TO EARLIER EDITIONS

Note: This book is a combined and updated edition of Volumes 1 and 2 of Roy Chudley's original *Construction Technology*.

Volume 1

In writing this book it is not my intention to present a comprehensive reference book on elementary building technology, since there are many excellent textbooks of this nature already in existence, which I urge all students to study.

My purpose, therefore, has been to prepare in concise note form, with ample illustrations, the basic knowledge the student should acquire in the first year of any building technology course of study.

To keep the book within a reasonable cost limit I have deliberately refrained from describing in depth what has been detailed in the drawings.

Building technology is an extensive but not necessarily exact subject. There are many ways of obtaining a satisfactory construction in building, but whichever method is used they are all based upon the same basic principles and it is these which are learnt in building technology.

The object of any building technology course is to give a good theoretical background to what is essentially a practical subject. With the knowledge acquired from such a course, coupled with observations of works in progress and any practical experience gained, a problem should be able to be tackled with confidence by the time the course of study has reached an advanced level.

Another aspect of a course of this nature is to give sufficient basic knowledge, over the whole field of building activities, to enable the technologist to hold and understand discussions with other related specialists.

Volume 2

The aim of this second volume is to provide a continuity of study with the contents of Volume 1, which dealt mainly with domestic building. The second year of a typical construction technology course will study further aspects of domestic construction and introduce the student to framing techniques and materials. The presentation is in the form of notes accompanied by ample illustrations with as little repetition between notes and drawings as possible. It is also written with the assumption that the reader has covered, understood and retained the technical knowledge contained in a basic first year course.

The depth of study presented in this volume is limited to the essential basic knowledge required for a second year construction technology course. The student is therefore urged to consult all other possible sources of reference to obtain a full and thorough understanding of the subject of construction technology.

Roy Chudley

PREFACE

Roy Chudley's *Construction Technology* was first published in four volumes, between 1973 and 1977. The material has since been continuously updated through numerous reprints and full second editions in 1987. The books have gained a world-wide readership, and their success – and their impact on construction education – is a tribute to Roy Chudley's experience in further and higher education and his talents as a skilled technologist, illustrator and writer.

As a former colleague, it has been a privilege to once again work with Roy, on this occasion revising his original work, and compiling the material into two books: *Construction Technology* and *Advanced Construction Technology*. The content forms a thorough study for all students of building, construction management, architecture, surveying and the many other related disciplines within the diverse construction profession.

The original presentation of comprehensive text matched by extensive illustration is retained. Changes in legislation, such as the Building and Construction Regulations, have been fully incorporated into the text; however, as much of the original work as possible has been purposely retained as it contains many relevant examples of existing construction. Additional material discusses the new developments and concepts of contemporary practice.

The two new volumes are complimentary, as many of the topics introduced here are further developed in *Advanced Construction Technology*. Together the books provide essential reading for all students aspiring to management, technologist and professional qualifications. They should be read alongside the current local building regulations and national standards, and where possible supplemented by direct experience in the workplace.

Roger Greeno
Guildford 1998

ACKNOWLEDGEMENTS

The revising author would like to thank James Newall and Brett Gilbert of Addison Wesley Longman: James for his enthusiasm and initiative to proceed with the books, and his invitation to participate; Brett for his untiring efforts in processing my manuscript, correlating this with Roy's original work and achieving a commendable presentation. I am especially grateful to Roy Chudley for his cooperation, trust and permission to work on his original manuscript, and to Colin Bassett, the General Editor of the earlier editions for his initial involvement.

Roger Greeno
Guildford 1998

We are grateful to the Building Research Establishment and The Stationery Office for permission to reproduce material from *BRE Digests* and various Acts, Regulations and Statutory Instruments.

Extracts from British Standards are reproduced with the permission of BSI. Complete copies can be obtained by post from BSI Customer Services, 389 Chiswick High Road, London W4 4AL; Telephone: 0181 996 7000.

INTRODUCTION

There are two general aspects to the construction of buildings:

1. Conventional or traditional methods.
2. Modern or industrialized methods.

Conventional or traditional methods are studied in the first two years of most construction courses, with the intention of forming a sound knowledge base before proceeding to studies of advanced techniques in the final years. There is nevertheless, an element of continuity and overlap between traditional and contemporary, and both are frequently deployed on the same building, e.g. traditional brick facing to a prefabricated steel-framed commercial building or to a factory-made timber-framed house.

Initial studies of building construction concentrate on the smaller type of structure, such as a domestic dwelling of one or two storeys built by labour-intensive traditional methods. Generally it is more economic to construct this type of building by these methods, unless large numbers of similar units are required on the same site. In these circumstances, economies of scale may justify factory manufactured, prefabricated elements of structure. These industrialized methods are usually a rationalized manufacturing process used to produce complete elements, i.e. floors, walls, roof frames, etc. in modules or standardized dimensional increments of 300 mm.

Very few building contractors in the UK and other developed countries directly employ many staff. They are therefore relatively small companies when compared to the capital value of the work they undertake. This is partly due to the variable economic fortunes of the construction industry and the need for flexibility. Hence, most practical aspects of building are contracted out to specialist subcontracting organizations, e.g. bricklayers, electricians, carpenters, etc., in response to the main contractor's work load. The main contractor is effectively a building management company that could be engaged on a variety of work, including major serial developments for the same client, maintenance work or after-care programmes, extensions to existing structures or possibly just small one-off projects.

It is essential that all students of building have an awareness of the variable methods of construction and application of materials in both traditional and industrial practice, in order to adapt their career pattern to the diverse expectations of the industry.

THE BUILDING TEAM

Building is essentially a team process in which each member has an important role to play. Figure 1 shows the organization structure of a typical team for a large project and the function of each member is outlined below.

Building owner   The client; the person or organization who finances and commissions the work. They directly or indirectly employ all other personnel, with particular responsibility for appointing the planning supervisor (usually the

3

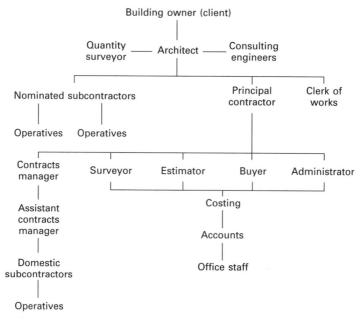

Figure 1   The building team

architect) and nominating the principal contractor – see Construction (Design and Management) Regulations 1994.

Architect   Engaged by the building owner as agent to design, advise and ensure that the project is kept within cost and complies with the design.

Clerk of works   Employed on large contracts as the architect's on-site representative. The main function is to liaise between architect and main contractor and to ensure that construction proceeds in accordance with the design. They can offer advice, but directives must be through the architect.

Quantity surveyor   Engaged to prepare cost evaluations and bills of quantities, check tenders, prepare interim valuations, effect cost controls and advise the architect on the cost of variations.

Consulting engineers   Engaged to advise and design on a variety of specialist installations, e.g. structural, services, security, etc. They are employed to develop that particular aspect of the design within the cost and physical parameters of the architect's brief.

Principal or main contractor   Employed by the client on the advice of the architect, by nomination or competitive tendering. They are required to administer the construction programme within the architect's direction.

Contract's manager or site agent   On large projects, the main contractor's representative on-site, with overall responsibility for ensuring that work proceeds effectively and efficiently, i.e. in accordance with the design specification and to time. Sometimes known as the general foreman, but this title is more appropriate on small to modest size contracts.

Surveyor   Employed by the main contractor to check work progress and assist the quantity surveyor in the preparation of interim valuations for stage payments and final accounts. May also be required to measure work done for bonus and subcontractor payments.

Estimator   Prepares unit rates for the pricing of tenders and carries out pre-tender investigations into the cost aspects of the proposed contract.

Buyer   Orders materials, obtains quotations for the supply of materials and services.

Accountant   Prepares and submits accounts to clients and makes payments to suppliers and subcontractors. May also have a costing department which would allocate the labour and material costs to each contract to assist with the preparation of accounts.

Administrator   Organizes the general clerical duties of the contractor's office for the preparation of contract documents and payment of salaries, subcontractors and suppliers invoices, insurances and all necessary correspondence.

Assistant contract manager   Often a trainee in the process of completing professional examinations. Assists with the general responsibility for administering site proceedings.

Nominated subcontractor   Engaged by the client or architect for specialist construction or installation work, e.g. lifts, air conditioning, etc.

Domestic subcontractor   Employed by the principal contractor to assist with the general construction, e.g. ground workers, bricklayers, etc.

Operatives   The main workforce on-site, includes craftsmen, apprentices and labourers.   '

The size of the building firm or the size of the contract will determine the composition of the construction team. For medium-sized contracts some of the above functions may be combined, i.e. that of the surveyor and estimator. Furthermore, many design and build practices have been created by combining the professional expertise of architect, builder and consultants. The objective is to improve communications and create better working relationships to provide the client with a more efficient and cost-effective service.

LEGISLATION

Many statutes made by Act of Parliament affect the construction of buildings and associated work. The most significant are:

1. The Health and Safety at Work, etc., Act 1974.
2. The Building Act 1984.

THE HEALTH AND SAFETY AT WORK, ETC., ACT 1974

This requires employers or their agents to implement an overall duty of care through a safe system of working. This applies to people in the workplace and others that could be affected by the work activity. The Act provides an all-embracing standard supporting a framework of statutory instruments or regulations administered by the Health and Safety Executive (HSE) through specialist inspectors operating from local offices. The inspectorate has powers to access premises, review a company's safety records, issue improvement and prohibition notices and, if necessary, effect prosecutions for non-compliance.

The following are the principal statutory instruments made under The Health and Safety at Work, etc., Act. They are legally binding on the client, architect and builder:

The Construction (Design and Management) Regulations 1994

Establishes mutual responsibility between client, designer and builder for matters pertaining to the health and safety of site personnel throughout the duration of a construction project. The client has prime responsibility for appointment of a 'planning supervisor' (usually the architect) and a 'principal contractor' (usually the main contractor) and to ensure that both are adequately resourced, competent and sufficiently informed of issues relating to the development.

The planning supervisor

Main responsibilities are to notify the HSE of the project details, to ensure co-operation between designers, to co-ordinate the design with regard to avoidance of undue risks, to prepare a pre-tender health and safety plan and manual for the client's use for advice and updates of health and safety issues.

The principal contractor

Responsibilities are mainly to ensure site personnel comply with the health and safety plan, to develop and update it throughout the duration of work and to exclude from site unauthorized and uninsured persons.

The Construction (Health, Safety and Welfare) Regulations 1996

Establishes objectives measured against assessment of risk, for the well-being of personnel on a building site throughout the duration of work. The main areas for assessment include: temporary timbering and other non-permanent support facilities, safety barriers to excavations, air quality in the workplace, safe use of doors, gates and other possible means of entrapment, defined traffic and pedestrian routes, safe means of access and egress to all workplaces including special considerations for scaffolds, and emergency lighting and power. Also, provision must be made for welfare facilities to include: sanitation, hot and cold water supply, first aid equipment/personnel, protective clothing, facilities to dry clothes and appropriate accommodation for meals. Implicit is good site management with regard for organization and planning.

Control of Substances Hazardous to Health Regulations 1994

Redresses the balance in favour of health, since the introduction of The Health and Safety at Work, etc., Act, has greater emphasis on safety. Manufacturing companies are obliged to monitor and declare health risks of their products and building contractors must provide operatives with protective clothing and/or a well-ventilated environment if required to use them. COSHH has promoted the removal of harmful substances from materials, e.g. toxins, irritants, solvents, etc., but finding substitutes is not always possible. Timber preservatives, welding fumes, dust from cement and plaster and insulating fibres are, as yet, a few of the unavoidable harmful constituents in building materials. Where these are applied, employers are obliged to monitor exposure levels, retain records, identify personnel who could be at risk and document the facilities provided for their protection.

The Manual Handling Operations Regulations 1992

Determines the employers' responsibility to ensure that employees under their control are not expected to undertake manual tasks that impose an undue risk of injury. The client, or their agent the main contractor, must assess all manual handling operations and reduce risk or injury or safety to the lowest practicable level. This will be manifest in the provision of work systems, such as cranes and hoists, that the employee is obliged to make full use of. Due regard must be applied to:

1. Tasks: space availability, manipulating distance, body movement, excess pulling and pushing and prolonged physical effort.
2. Load: unwieldy or bulky, eccentric, excessive, liability to shift, temperature and sharpness of finish.
3. Work environment: surface finish (slippery, uneven or variable), lighting, ventilation and temperature variation.
4. Individual capacity: unusual weight or dimension, health problems/limitations, special training and provision of personal protective clothing.

THE BUILDING ACT 1984

This is a consolidating Act of primary legislation relating to building work. It contains enabling powers and a means for the Secretary of State for the Environment to produce building regulations for the outline purpose of:

1. Maintaining the health, well-being and convenience for people using and relating to buildings.
2. Promoting the comfort of building occupants, with due regard for the conservation of fuel and efficient use of energy within the structure.
3. Preventing the misuse of water, excessive consumption and contamination of supply.

The building regulations or statutory instruments currently in place under the Building Act are:

- The Building Regulations applicable to England and Wales.
- The Building (Approved Inspectors, etc.) Regulations.
- The Building (Prescribed Fees) Regulations.
- The Building (Inner London) Regulations.
- The Building (Disabled People) Regulations.

The Building Regulations contain minimum performance standards expected of contemporary buildings. They are supported by a series of Approved Documents which are not mandatory, but give practical guidance on compliance with the requirements of the regulations. This guidance often incorporates British Standard, Building Research Establishment, British Board of Agrément and other authoritive references.

Control of the Building Regulations is vested in the local council authority. However, a developer/builder can opt for private certification whereby the developer and an approved inspector jointly serve an initial notice on the local authority. This describes the proposed works, which the local authority can reject (if they have justification) within ten days of application. With this option, the responsibility for inspecting plans, quality of work, site supervision and certification of satisfactory completion, rests with the approved inspector as defined in The Building (Approved Inspectors, etc.) Regulations.

Building legislation is regionally divided throughout the UK, each area having its own regulations:

- The Building Regulations applicable to England and Wales.
- The Building Standards (Scotland) Regulations.
- The Building Regulations (Northern Ireland).
- The Building (Inner London) Regulations.

There are also statutory provisions for the Channel Islands, Isle of Man and the Republic of Ireland.

BRITISH STANDARDS

These function as support documents to building design and practice. Many are endorsed by the building regulations and provide recommendations for minimum material and practice specifications. They are published in four possible formats:

1. **Codes of practice (BS Code of Practice for . . .)**   These are guides to good practice in particular areas of construction activity, e.g. scaffolding and structural steelwork.
2. **Standard specifications (BS)**   These are specifications appropriate to materials and components such as bricks and windows. Products complying may be endorsed with the Institution's kitemark.
3. **Draft for development (DD)**   These are issued instead of a Code of Practice or Standard Specification where there is insufficient data or information to make a firm or positive recommendation. They have a maximum life of five years, during which sufficient data may be accumulated to upgrade the draft to a Standard specification.
4. **Published document (PD)**   These are publications which are difficult to locate in any of the preceding categories.

Codes of practice and standard specifications are compiled by specialist committees of professional interest in the particular subject. It must be remembered that these codes and specifications are only recommendations. However, many are used in the Building Regulations, Approved Documents as a minimum acceptable standard for specific applications. Copies of the codes and specifications can be obtained from the British Standards Institution (Customer Services), 389 Chiswick High Road, London, W4 4AL.

AGRÉMENT CERTIFICATES

These are administered by the British Board of Agrément, Bucknalls Lane, Garston, Watford WD2 7NG. Their purpose is to establish the quality and suitability of new products and innovations, not covered by established performance documents such as British (BSI) or European Standards (CEN). The board assesses, examines and tests materials and products with the aid of the Building Research Establishment and other research centres, to produce reports. These reports are known by the acronym – MOATS (Methods Of Assessment and Testing). Products satisfying critical examination relative to their declared performance are issued with an Agrément Certificate.

INTERNATIONAL STANDARDS

Many British Standards have been harmonized with the dimensional and material specifications of European continental products and vice-versa, e.g. BS EN 196–2:

1995: *Chemical analysis of cement*, has replaced BS 4550: Part 2: 1970: *Methods of testing cement; chemical tests*. European standards are administered by the Comité Européen de Normalisation (CEN), which incorporates bodies such as the British Standards Institution (BSI) from member states.

The International Standards Organization also produce documents compatible with British Standards. These are prefixed ISO, e.g. BS EN ISO 9000–4: *Quality and quality assurance standards*, complements BS 5750: *Quality systems*.

METRICATION

Since the building industry adopted the metric system of measurement during the early 1970s, there has been a great deal of modification and rationalization as designers and manufacturers have adapted themselves to the new units.

Most notable has been product developments co-ordinated about the preferred dimension of 300 mm. The second preference is 100 mm, with subdivisions of 50 mm and 25 mm as third and fourth preferences respectively. Many manufacturers now base their products on the 100 mm module and co-ordinate their components within a three-dimensional framework of 100 mm cubes. This provides product compatibility and harmonization with international markets that are also metric and dimensionally co-ordinated. Kitchen fitments provide a good example of successful development, with unit depths of 600 mm and widths in 300, 600, 1,200 mm and so on. Less successful has been the move to metric bricks in lengths of 200 or 300 mm × 100 mm wide × 75 mm high. They are just not compatible with existing buildings and bricklayers have found them less suited to the hand.

Although the metric system is now well established, it should be remembered that a large proportion of the nation's buildings were designed and constructed using imperial units of measurement which are incongruous with their metric replacements. Many building materials and components will continue to be produced in imperial dimensions for convenience, maintenance, replacement and/or refurbishment, and where appropriate some manufacturers provide imperial–metric adaptors, e.g. plumbing fittings. The prohibitive capital cost of replacing manufacturing plant such as brick presses and timber converters will preserve traditional imperial dimensions for some time.

PART 1

SITE AND TEMPORARY WORKS

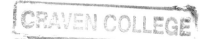

SITE WORKS AND SETTING OUT

When a builder is given possession of a building site the contractor will have been provided with the site layout plan and the detail drawings necessary for him to construct the building(s). Under most forms of building contract it is the builder's responsibility to see that the setting out is accurate.

The site having been taken over, the task of preparing for and setting out the building can be commenced. These operations can be grouped under three headings:

- Clearing the site.
- Setting out the building.
- Establishing a datum level.

CLEARING THE SITE

This may involve the demolition of existing buildings, the grubbing out of bushes and trees and the removal of soil to reduce levels. Demolition is a skilled occupation and should only be tackled by an experienced demolition contractor. The removal of trees can be carried out by manual or mechanical means. The removal of large trees should be left to the specialist contractors.

Building Regulation C1, 'The ground to be covered by the building shall be reasonably free from vegetable matter.' This is in effect to sterilize the ground since the top 300 mm or so will contain plant life and decaying vegetation. This means that the topsoil is easily compressed and would be unsuitable for foundations. Topsoil is valuable as a dressing for gardens and will be retained for reinstatement when the site is landscaped. The method chosen for conducting the site clearance work will be determined by the scale of development and consideration for any adjacent buildings.

SETTING OUT THE SITE

The first task is to establish a base line from which the whole of the building can be set out. The position of this line must be clearly marked on-site so that it can

13

Figure 1.1.1 Setting out and checking methods

Trench width

Wall width

150 × 38 boards 450 wider than trench

Position of trench plumbed down

Nails positioning trench and walls (alternative sawcuts)

Setting out lines

50 × 50 pointed posts driven into ground 450 to 600 deep height above ground 150 to 600

Figure 1.1.2   Typical profile board

be re-established at any time. For on-site measuring a steel tape should be used (30 metres would be a suitable length). Linen and plastic-coated tapes are also available. The disadvantage with linen tapes is that they are liable to stretch.

After the base line has been set out, marked and checked the main lines of the building can be set out, each corner being marked with a stout peg. A check should now be made of the setting-out lines for right angles and correct lengths. There are several methods of checking if a right angle has been established and in fact the setting out would have been carried out by one of these methods. A check must still be made and it is advisable to check by a different method to that used for the setting out. The setting-out procedure and the methods of checking the right angles are illustrated in Fig. 1.1.1.

After the setting out of the main building lines has been completed and checked, profile boards are set up as shown in Fig. 1.1.2. These are set up clear of the foundation trench positions to locate the trench, foundations and walls. Profile boards are required at all trench and wall intersections.

ESTABLISHING A DATUM LEVEL

It is important that all levels in a building are taken from a fixed point called a 'datum'. This point should now be established; wherever possible this should be related to an ordnance benchmark. This is an arrow with a horizontal mark above

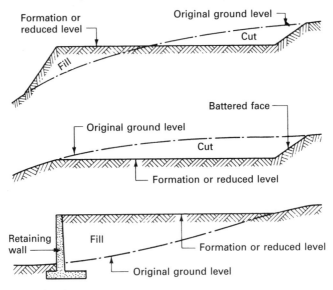

Figure 1.1.3   Sloping sites

the arrow, the centreline of the horizontal being the actual level indicated on an ordnance survey map. Benchmarks are found cut or let into the sides of walls and buildings. Where there are no benchmarks on or near the site a suitable datum must be established. A site datum or temporary benchmark could be a post set in concrete or a concrete plinth set up on site.

SLOPING SITES

Very few sites are level and therefore before any building work can be commenced the area covered by the building must be levelled. In building terms this operation is called reducing levels. Three methods can be used and it is the most economical which is usually employed.

1. **Cut and fill**   The usual method because, if properly carried out, the amount of cut will equal the amount of fill.
2. **Cut**   This method has the advantage of giving undisturbed soil over the whole of the site but has the disadvantage of the cost of removing the spoil from the site.
3. **Fill**   A method not to be recommended because, if the building is sited on the filled area, either deep foundations would be needed or the risk of settlement at a later stage would have to be accepted. The amount of fill should never exceed a depth of 600 mm.

The principles of the above methods are shown in Fig. 1.1.3.

ACCOMMODATION, STORAGE AND SECURITY

The activities and the temporary nature of a building site do not generally justify the provision of permanent buildings for staff accommodation or for the storage of materials. It is, however, within the builder's interest to provide the best facilities which are economically possible for any particular contract; this should promote good relationships between management and staff, it should also reduce the loss of materials due to theft, accidental damage and vandalism. The better the facilities and amenities provided on a building site the greater will be the contentment of the site staff which will ultimately lead to higher productivity.

ACCOMMODATION

The Construction (Health, Safety and Welfare) Regulations 1996 is a statutory instrument which establishes objectives for accommodation and facilities for staff to be provided on sites throughout the construction industry. Requirements will vary with regard to the number of personnel on site and in some cases the anticipated duration of the contract. An extract from the 1966 forerunner to the current regulations is shown in Table 1.2.1, which indicates absolute minimum standards for guidance only.

Units of staff accommodation usually come in one of two forms:

1. Sectional timber huts.
2. Mobile caravans or cabins.

Sectional timber huts are prefabricated to allow for ease of dismantling and assembly to facilitate the re-use on other sites. Huts of this nature should be designed, constructed and maintained with the same care as permanent buildings to ensure their use for many years on a number of different contracts. A well-designed sectional hut should permit the addition of more bays to increase the modular size by length and/or width.

The anticipated use of each hut will govern the construction and facilities required. Offices need to be weatherproof, heated, insulated to conserve the heat, have some form of artificial lighting, and be equipped with furniture such as desks,

17

Table 1.2.1 The Construction (Health and Welfare) Regulations 1996

No. of persons employed by contractor on site	0 5 10	20 25	40	50	100
First aid	Box to be clearly marked and in charge of named person.	First-aid boxes	First-aid boxes and person trained in first aid →		
Stretcher ambulance		Stretcher provided. Local health authority informed of site, work and completion date. If no 'phone or radio, ambulance kept ready →			
First-aid room	To be used only for treatment and in charge of trained person.		Where number of persons on site exceeds 250 each employer of more than 40 persons must provide first-aid room →		
Shelter and clothing	All persons to have shelter and place for depositing clothing.	Adequate means of warming themselves and drying wet clothing	Where possible, means of warming themselves and drying wet clothing →		
Meals room	All persons to have drinking water provided and facilities for boiling water and eating meals.	Facilities for heating food if hot meals are not available on site →			
Washing facilities	All persons on site for more than 4 hours to have washing facilities.	Where work is likely to last 6 weeks hot and cold or warm water, soap and towel provided			Where work is likely to last 12 months 4 washplaces plus 1 for every 35 persons more than 100 →
Sanitary facilities	To be maintained and kept clean – provision to be made for lighting.	1 convenience for every 25 persons		1 convenience for every 35 persons →	

Notes

- Washing facilities to be close to meals room.
- Protective clothing to be provided where person is required to work in inclement weather.
- Subcontractors may use the facilities provided by another contractor and for the purpose of these regulations their workforce on site is included in the total workforce on site.

Typical section

Plan

Figure 1.2.1 Typical prefabricated timber site office

work tops, plan chests and chairs to suit the office activities; a typical timber sectional site office detail is shown in Fig. 1.2.1. The same basic construction can be used for all other units of accommodation such as meals rooms and toilets equipped with the facilities shown in Table 1.2.1.

Caravans and mobile cabins are available in a wide variety of sizes, styles and applications. The construction is generally of a ply clad timber frame suitably insulated and decorated; they are made to a modular system so that by using special connection units any reasonable plan size and shape is possible. The caravans and cabins are fully equipped with all the necessary furniture, lights and heating units. The toilets are supplied with all the necessary sanitary fittings and plumbing which can be connected to site services or be self-contained. Transportation of caravans or cabins can be on any suitable vehicle; caravans can be towed whereas special transporter trailers are available for cabins. Whichever method is used the time taken to load, offload and position on site is considerably shorter than the time required to dismantle, transport and reassemble a sectional timber hut, but the initial capital outlay is higher. Typical examples of units of accommodation are shown in Fig. 1.2.2.

STORAGE

The type of storage facilities required of any particular material will depend upon the following factors:

- Durability – will it need protection from the elements?
- Vulnerability to damage.
- Vulnerability to theft.

Cement, plaster and lime supplied in bag form require a dry store free from draughts which can bring in moist air and may cause an air set of material. These materials should not be stored for long periods on site, therefore provision should be made for rotational use so that the material being used comes from the older stock.

Aggregates such as sand and ballast require a clean firm base to ensure that foreign matter is not included when extracting materials from the base of the stock pile. Different materials and grades must be kept separated so that the ultimate mix batches are consistent in quality and texture. Care must be taken, by careful supervision, to ensure that the stock piles are not used as a rubbish tip. If the storage piles are exposed to the elements a careful watch should be kept on the moisture content; if this rises it must be allowed to drain after heavy rain or alternatively the water/cement ratio of the mix can be adjusted.

Bricks and blocks should be stacked in stable piles on a level and well-drained surface in a position where double handling is reduced to a minimum. Facing bricks and light-coloured bricks can become discoloured by atmospheric pollution and/or adverse weather conditions; in these situations the brick stacks should be covered with tarpaulin or polythene sheeting adequately secured to prevent dislodgement. Blocks, being less dense than bricks, should be stacked to allow air movement around them and should always be covered with a suitable sheet material.

Sectional store hut

Cabin toilet unit

Typical layout

Mobile office

Figure 1.2.2 Typical site accommodation units

Roof tiles have a greater resistance to load when it is imposed on the edge; for this reason tiles should be stacked on edge and in pairs, head to tail, to give protection to the nibs. An ideal tile stack would be five to seven rows high with end tiles laid flat to provide an abutment. Tile fittings such as ridge and hip tiles should be kept separate and if possible placed on end.

Drainage goods, like tiles, may be stored in an open compound; they should be stacked with their barrels horizontal and laid with spigots and sockets alternately reversed or placed in layers with the spigots and sockets reversed in alternate layers. Plain ended clay pipes are delivered to site on pallets with pipes separated by timber battens recessed to suit the pipe profile. Pipes should be stored in this way until they are required. Fittings should be kept separate and those like gullies, which can hold water, should be placed upside down.

Timber is a hygroscopic material and therefore to prevent undue moisture movement it should be stored in such a manner that its moisture content remains fairly constant. A rack of scaffold tubulars with a sheet roof covering makes an ideal timber store. The various section sizes allow good air flow around the timber and the roof provides protection from the rain and snow.

Ironmongery, hand tools and paint are some of the most vulnerable items on a building site. Small items such as locks, power drills and cans of paint should be kept in a locked hut and only issued against an authorized stores requisition. Large items like baths can be kept in the compound and suitably protected; it is also good practice only to issue materials from the compound against a requisition order.

SECURITY AND PROTECTION

FENCING

A building site and the compound can be given a degree of protection by surrounding with a fence. The fence fulfils two functions:

1. Defines the limit of the site or compound.
2. Acts as a deterrent to the would-be trespasser or thief.

A fence can be constructed to provide a physical barrier of solid construction or a visual barrier of open work construction. If the site is to be fenced as part of the contract it may be advantageous to carry out this work at the beginning of the site operations. The type of fencing chosen will depend upon the degree of security required, cost implications, type of neighbourhood and duration of contract.

A security fence around the site or compound should be at least 1.800 m high above the ground and include the minimum number of access points which should have a lockable barrier or gate. Standard fences are made in accordance with the recommendations of BS 1722 which covers ten forms of fencing giving suitable methods for both visual and physical barriers; typical examples are shown in Fig. 1.2.3.

Cleft chestnut pale fencing

Chain link fence with concrete posts

Close boarded fence with concrete posts

Figure 1.2.3 Typical fencing details

Panels constructed of 100 × 50 studs, rails and noggins covered with close boarding or sheeting

Night warning lights

1.200 min.

100 × 50 braces

1.800 to 2.100

Guard rail

300 × 300 timber fender painted white

900

50 × 50 fixing stakes

Paving slabs removed

600

Footway of t & g boards on 125 × 50 joists at 450 c/c

Typical free-standing vertical hoarding

Close boarded or sheeted deck

100 × 50 timber supports at 600 c/c anchored to existing floor

Struts at 600 c/c required if cantilever exceeds 1.800

Hole through existing wall for timber supports

Note: If hoarding is over a public footpath it should be lined with polythene or similar sheeting

Typical fan hoarding

Figure 1.2.4   Timber hoardings

HOARDINGS

These are close boarded fences or barriers erected adjacent to a highway or public footpath to prevent unauthorized persons obtaining access to the site and to provide a degree of protection for the public from the dust and noise associated with building operations. Under Sections 172 and 173 of the Highways Act 1980 it is necessary to obtain written permission from the local authority to erect a hoarding. The permission, which is in the form of a licence, sets out the conditions and gives details of duration, provision of footway for the public and the need for lighting during the hours of darkness.

Two forms of hoarding are in common use:

1. Vertical hoardings.
2. Fan hoardings.

The vertical hoardings consist of a series of closed boarded panels securely fixed to resist wind loads and accidental impact loads. It can be free standing or fixed by stays to the external walls of an existing building (see Fig. 1.2.4).

The Construction (Health, Safety and Welfare) Regulations 1996 requires protection to be given to persons from falling objects. A fan hoarding fulfils this function by being placed at a level above the normal traffic height and arranged in such a manner that any falling débris is directed back towards the building or scaffold (see Fig. 1.2.4).

SUBSOIL DRAINAGE

BUILDING REGULATION C2 Resistance to moisture

Subsoil drainage shall be provided if it is needed to avoid:

- The passage of ground moisture to the interior of the building.
- Damage to the fabric of the building.

The ideal site (see Fig. 1.3.1) will not require any treatment but sites with a high water table will require some form of subsoil drainage. The water table is the level at which water occurs naturally below the ground and this level will vary with the seasonal changes.

The object of subsoil drainage is to lower the water table to a level such that it will comply with the above building regulation, i.e. not rise to within 0.25 m of the lowest floor of a building. It also has the advantage of improving the stability of the ground, lowering the humidity of the site and improving its horticultural properties.

MATERIALS

The pipes used in subsoil drainage are usually dry jointed and are either porous or perforated pipes. The porous pipes absorb the water through their walls and thus

Figure 1.3.1   The ideal site

26

Herringbone   **Grid iron**   **Moat**

(Branch drains spacing 6.000 to 10.000 – max length 30.000)

Figure 1.3.2   Subsoil drainage systems and drains

keep out the fine particles of soil or silt, whereas perforated pipes, which are laid with the perforations at the base, allow the water to rise into the pipe leaving any silt behind.

Suitable pipes

- Perforated clayware – BS EN 295–5.
- Porous concrete – BS 5911: Part 114.
- Clayware field pipes – BS 1196.
- Profiled and slotted polypropylene or uPVC – BS 4962.
- Perforated uPVC – BS 4660.

DRAINAGE LAYOUTS

The pipes are arranged in a pattern to cover as much of the site as is necessary. Typical arrangements are shown on the plans in Fig. 1.3.2. Water will naturally flow towards the easy passage provided by the drainage runs. The system is

Figure 1.3.3   Outfall to stream or river

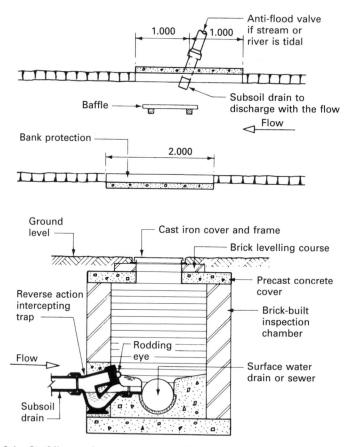

Figure 1.3.4   Outfall to surface water sewer or drain

terminated at a suitable outfall such as a river, stream or surface water sewer. In all cases permission must be obtained before discharging a subsoil system. The banks of streams and rivers will need protection against the turbulence set up by the discharge and if the stream is narrow the opposite bank may also need protection (see Fig. 1.3.3). If discharge is into a tidal river or stream precautions should be taken to ensure the system will not work in reverse by providing an outlet for the rising tide. On large schemes sediment chambers or catch pits are sometimes included to trap some of the silt which is the chief cause of blockages in subsoil drainage work. The construction of a catch pit is similar to the manhole shown in Fig. 1.3.4 except that in a catch pit the inlet and outlet are at a high level; this interrupts the flow of subsoil water in the drains and enables some of the silt to settle on the base of the catch pit. The collected silt in the catch pit must be removed at regular intervals.

EXCAVATIONS AND TIMBERING

Before a foundation can be laid it is necessary to excavate a trench of the required depth and width. On small contracts such as house extensions this is effectively carried out by hand, but on large works it will be more economic to use some form of mechanical excavator. The general procedure for the excavation of foundation trenches is illustrated in Fig. 1.4.1.

Figure 1.4.1   Trench excavations

Figure 1.4.2 Typical timbering in hard soils

TIMBERING

This is a term used to cover temporary supports to the sides of excavations and is sometimes called planking and strutting. The sides of some excavations will need support to:

■ Protect the operatives while working in the excavation.
■ Keep the excavation open by acting as a retaining wall to the sides of the trench.

The type and amount of timbering required will depend upon the depth and nature of the subsoil. Over a short period many soils may not require any timbering but weather conditions, depth, type of soil and duration of the operations must all be taken into account and each excavation must be assessed separately.
   Suitable timbers for this work are:

■ Scots pine.
■ Baltic redwood.
■ Baltic whitewood.
■ Douglas fir.
■ Larch.
■ Hemlock.

Typical details of timbering to trenches are shown in Figs 1.4.2–1.4.6.

THE CONSTRUCTION (HEALTH, SAFETY AND WELFARE) REGULATIONS 1996

This document establishes objectives for employers, the self-employed and employees, to ensure safe working and support in excavations.

Figure 1.4.3   Typical timbering in firm soils

Figure 1.4.4   Typical timbering in dry loose soils

Figure 1.4.5 Typical timbering in loose wet soils

Size no.	Minimum length (mm)	Maximum length (mm)
0	300	480
1	480	680
2	680	1.060
3	1.060	1.670

Figure 1.4.6 Adjustable metal struts – BS 4074

Regulation 5 – Safe places of work

This is a general requirement applying not least to work in the ground. It places an obligation on all involved, to assess the risks and to ensure reasonably practicable means to safeguard work in excavations.

Regulations 12 and 13 – Excavations

Timber or other suitable material must be provided and used to prevent danger from a fall or dislodgement of materials forming the sides of an excavation.

Underground services must be foreseen (if possible), located, identified and assessed for risk to operatives working in excavations. Positive action is necessary to eliminate any risk and to prevent injury.

Deep excavations, tunnels, cofferdams and caissons require special consideration and must be correctly designed, constructed and maintained with regard for prevailing conditions. (See Chapter 3 of *Advanced Construction Technology*.)

Regulations 28, 29 and 30 – Training, inspection and reports

This requires excavation support to be installed or supervised by adequately trained, suitably knowledgable and experienced personnel. Prior to general access being given, an inspection of excavations by a competent person (usually the main contractor's safety supervisor) must be made. That person must ensure that work can proceed safely. Following inspection, a written report must be filed for presentation if requested by the Health and Safety Executive. Further documented inspections should be undertaken daily or at the beginning of each shift or after severe weather conditions. Changes to the trench format and unexpected falls of earth should also be monitored.

BUILDING REGULATIONS

Notice of commencement and completion of certain stages of work.

Building Regulation 14 requires that the building control office of the local authority is notified by a person carrying out building work prior to commencement and at specific stages during construction work. The notice should be given in writing or by such means as may be agreed with the local authority.

Notices of the stages when statutory inspections are required under this regulation occur as follows:

- Commencement   48 h
- Foundation excavation   24 h
- Foundation concrete   24 h
- Oversite preparation, before concreting   24 h
- Damp-proof course   24 h
- Drains before backfilling (foul and rainwater)   24 h

- Drain test after covering   5 days after
- Occupation before completion   5 days before
- Completion   within 5 days

With the exception of notice of commencement, drain testing and completion, the amount of time required before progressing with other parts of the work is one day. Weekends and bank holidays are excluded.

SCAFFOLDING

A scaffold is a temporary structure from which persons can gain access to a place of work in order to carry out building operations, it includes any working platforms, ladders and guard rails. Basically there are two forms of scaffolding:

- Putlog scaffolds.
- Independent scaffolds.

PUTLOG SCAFFOLDS

This form of scaffolding consists of a single row of uprights or standards set away from the wall at a distance which will accommodate the required width of the working platform. The standards are joined together with horizontal members called ledgers and are tied to the building with cross members called putlogs. The scaffold is erected as the building rises and is mostly used for buildings of traditional brick construction (see Fig. 1.5.1).

INDEPENDENT SCAFFOLDS

An independent scaffold has two rows of standards which are tied by cross members called transoms. This form of scaffold does not rely upon the building for support and is therefore suitable for use in conjunction with framed structures (see Fig. 1.5.2).

Every scaffold should be securely tied to the building at intervals of approximately 3.600 m vertically and 6.000 m horizontally. This can be achieved by using a horizontal tube called a bridle bearing on the inside of the wall and across a window opening with cross members connected to it (see Fig. 1.5.1); alternatively a tube with a reveal pin in the opening can provide a connection point for the cross members (see Fig. 1.5.2). If suitable openings are not available then the scaffold should be strutted from the ground using raking tubes inclined towards the building.

Figure 1.5.1 Typical tubular steel putlog scaffold

MATERIALS

Scaffolding can be of:

- Tubular steel.
- Tubular aluminium alloy.
- Timber.

TUBULAR STEEL

British Standard 1139 gives recommendations for both welded and seamless steel tubes of 48 mm outside diameter with a nominal 38 mm bore diameter. Steel tubes

Wall under construction

300

Inner board

Boarded
working
platform

Transom

Ledger

Cross
brace

Head
plate

Strut

Transom

Ledger

Sole
plate

Reveal
pin

1.350 to 1.500

Cross
brace

Inner
standards
at 2.000 c/c

Base plate

Guard rail

Mesh brick guard

Intermediate guard rail

Toe board clip

Toe board

Ledger

Ladder secured
by rope or clips
to top ledger

Outer
standards
at 2.000 c/c

Ledger

Sole
plate

Ladder
fixing
stake

Note: Not more than 50% of ties should be reveal ties

Figure 1.5.2   Typical tubular steel independent scaffold

can be obtained galvanized (to guard against corrosion); ungalvanized tubes will require special care such as painting, varnishing or an oil bath after use. Steel tubes are nearly three times heavier than comparable aluminium alloy tubes but are far stronger and since their deflection is approximately one-third of aluminium alloy tubes, longer spans can be used.

ALUMINIUM ALLOY

Seamless tubes of aluminium alloy with a 48 mm outside diameter are specified in BS 1139 for metal scaffolding. No protective treatment is required unless they are to be used in contact with materials such as damp lime, wet cement and sea water, which can cause corrosion of the aluminium alloy tubes. A suitable protective treatment would be to coat the tubes with bitumastic paint before use.

TIMBER

The use of timber as a temporary structure in the form of a scaffold is now rarely encountered in the UK, although it is extensively used in the developing world. The timber used is structural quality softwood in either putlog or independent format. The members are lashed together with wire or rope instead of the coupling fittings used with metal scaffolds.

SCAFFOLD BOARDS

These are usually boards of softwood timber complying with the recommendations of BS 2482 used to form the working platform at the required level. They should be formed out of specified softwoods of 225×38 section and not exceeding 4.800 m in length. To prevent the ends from splitting they should be end bound with not less than 25 mm wide $\times 0.9$ mm galvanized hoop iron extending at least 150 mm along each edge and fixed with a minimum of two fixings to each end. The strength of the boards should be such that they can support a uniformly distributed load of 6.7 kN/m^2 when supported at 1.200 m centres.

SCAFFOLD FITTINGS

Fittings of either steel or aluminium alloy are covered by the same British Standard as quoted above for the tubes. They can usually be used in conjunction with either tubular metal unless specified differently by the manufacturer. The major fittings used in metal scaffolding are:

- **Double coupler**   The only real loadbearing fitting used in scaffolding and is used to join ledgers to standards.
- **Swivel coupler**   Composed of two single couplers riveted together so that it is possible to rotate them and use them for connecting two scaffold tubes at any angle.
- **Putlog coupler**   Used solely for fixing putlogs or transoms to the horizontal ledgers.
- **Base plate**   A square plate with a central locating spigot used to distribute the load from the foot of a standard on to a sole plate or firm ground. Base plates can also be obtained with a threaded spigot and nut for use on sloping sites to make up variations in levels.
- **Split joint pin**   A connection fitting used to joint scaffold tubes end to end. A centre bolt expands the two segments which grip on the bore of the tubes.
- **Reveal pin**   Fits into the end of a tube to form an adjustable strut.
- **Putlog end**   A flat plate which fits on the end of a scaffold tube to convert it into a putlog.

Typical examples of the above fittings are shown in Fig. 1.5.3.

Figure 1.5.3   Typical steel scaffold fittings

Standards, putlogs and transoms

Stability of scaffolds

Platforms, gangways and runs

Boards in working platforms

Figure 1.5.4 Scaffolds and construction regulations – 1

Clear passage way for
persons or materials

Deposit of materials

Widths of working platforms for putlog and independent scaffolds

**Guard rails and toe boards for
putlog and independent scaffolds**

Use of ladders

Figure 1.5.5   Scaffolds and construction regulations – 2

THE CONSTRUCTION (HEALTH, SAFETY AND WELFARE) REGULATIONS 1996

This statutory instrument is designed to ensure that suitable and sufficient safe access to and egress from every place at which any person at any time works is provided and properly maintained. Scaffolds and ladders are covered by this document which sets objective requirements for materials, maintenance, inspection and construction of these working places. The main constructional requirements of these regulations are illustrated in Figs 1.5.4 and 1.5.5. Regulations 28, 29 and 30 determine the need for operatives' training, supervision and inspection of work, plus written reports by a competent person to support each inspection. Supervision of scaffold erection and progress/safety reports are undertaken by a suitably experienced and qualified person. This normally occurs:

- Within the preceding seven days.
- After adverse weather conditions which may have affected the scaffold's strength or stability.
- Whenever alterations or additions are made to the scaffold.

Records of all such inspections must be kept in accordance with the regulations and must provide the following information:

- Location and description of scaffold.
- Date of inspection.
- Result of inspection, stating the condition of scaffold.
- Signature and office of the person making the inspection.

The importance of providing a safe and reliable scaffold from which to undertake building work cannot be overemphasized. Badly assembled and neglected scaffolds have been a significant contributory factor to the high accident rate associated with the construction industry. The Health and Safety Executive (HSE) are promoting instructional literature and training programmes for scaffold erectors, as lack of formal training has been identified as the source of many site accidents.

PART 2

SUBSTRUCTURE

TRENCH AND BASEMENT EXCAVATION

Excavations may be classified as shallow, medium or deep in the following manner:

- Shallow – up to 1.500 m deep.
- Medium – 1.500 to 3.000 m deep.
- Deep – over 3.000 m deep.

The method of excavation and timbering to be used in any particular case will depend upon a number of factors:

1. Nature of the subsoil can determine the type of plant or hand tools required and the amount of timbering necessary.
2. Purpose of the excavation can determine minimum widths, minimum depths and the placing of support members to give a reasonable working space within the excavation.
3. Presence of ground water may necessitate the need for interlocking timbering, sump pits and pumps; large quantities of ground water may prompt the use of de-watering techniques.
4. Position of the excavation may impose certain restrictions such as the need for a licence or wayleave, highway authority or police requirements when excavating in a public road.
5. Non-availability of the right type of plant for bulk excavation may mean a different method must be used.
6. Presence of a large number of services may restrict the use of machinery to such an extent that it becomes uneconomic.
7. The disposal of the excavated spoil may restrict the choice of plant due to the load and unload cycle not keeping pace with the machine output.

SAFETY

As previously described in Chapter 1.1.4, safety requirements in subsurface excavations, including basements, are strictly enforced through the Construction (Health, Safety and Welfare) Regulations 1996. Additionally, a suitable fence or barrier must be provided to the sides of excavations. For small pits, such as manholes and other limited excavation, the void may be covered. Methods of providing a suitable barrier are shown in Fig. 2.1.1. Materials must not be placed near the edge of any excavation, nor must plant be placed or moved near excavations so that persons working in the excavation are endangered.

TRENCH EXCAVATIONS

Long narrow trenches in firm soil may be excavated to the full depth by mechanical excavators enabling the support timbering to be placed in one continuous operation. Weak and waterlogged ground must be supported before excavation commences by driving timber runners or steel trench sheeting to a position below the formation level or by a drive and dig procedure. In the latter method the runners can be driven to a reasonable depth of approximately 1.500 m followed by an excavation cut of 1.200 m and then the operation repeated until the required level has been reached; this will make the driving of the runners easier and enable a smaller driving appliance to be used.

In medium depth trenches different soil conditions are very often encountered throughout the depth of the excavation and therefore the method of timbering must be changed to suit the new soil conditions; a typical example of trench timbering in these circumstances is shown in Fig. 2.1.2.

Hand trimming is usually required in the trench bottom to form an accurate line and level; this process is called bottoming of trenches. Approximately 150 mm should be allowed for trimming by hand and it is advisable to cover the trimmed surface with hardcore to protect the soil at formation level from being disturbed or drying out and shrinking.

BASEMENT EXCAVATIONS

There are three methods which can be used for excavating a large pit or basement:

1. Complete excavation with sloping sides.
2. Complete excavation with timbered sides.
3. Perimeter trench method.

Excavation for a basement on an open site can be carried out by cutting the perimeter back to the natural angle of repose of the soil. This method requires sufficient site space around the intended structure for the over-excavation. No timbering is required but the savings on the temporary support work must pay for the over-excavation and consequent increase in volume of backfilling to be an economic method.

In firm soils where poling boards can be placed after excavation an economic method is to excavate the bulk of the pit and then trim the perimeter, placing

Brightly coloured plastic tape tied to metal stakes at 2 m max. spacing. (Proprietory interlocking rail and post system is preferred)

Plastic hazard cones secured by stakes

450 min.

600 to 900

Timbering

Close boarding, sheeting or runners left projecting above ground level

600 to 900

Timbering

Continuous rail fixed to the top of projecting poling boards

600 to 900

Timbering

Continuous tubular rail

450 min.

600 to 900

Timbering

Tripods of short scaffold tubes at 1.800 c/c

Figure 2.1.1 Barriers to excavations

Figure 2.1.2   Trench timbering in different soils

the poling boards with their raking struts in position as the work proceeds. Alternatively the base slab could be cast before the perimeter trimming takes place and the rakers anchored to its edge or side (see Fig. 2.1.3).

The perimeter trench method is used where weak soils are encountered; a trench wide enough to enable the retaining walls to be constructed is excavated around the perimeter of the site and timbered according to the soil conditions. The permanent

200 × 38 poling boards at centres to suit soil conditions

150 × 150 walings

150 × 150 puncheons at 2.000 c/c

150 × 150 raking struts at 2.000 c/c

Bulk excavation line

200 × 38 binding to both sides

w.i. dog

225 × 75 sole plate

Platform or grillage out of 200 × 75 timbers

150 × 75 sole plate

200 × 38 poling boards at centres to suit soil conditions

150 × 150 puncheons at 2.000 c/c

150 × 150 walings

200 × 38 binding to both sides

150 × 150 raking struts at 2.000 c/c

150 × 150 plate fixed to slab

225 × 75 sole plate

Basement slab cast before perimeter excavation and timbering

Alternative timbering methods for complete excavation

Trench timbering: see Fig. 2.1.2

Timbering rearranged as wall is cast

Retaining wall

Dumpling of unexcavated soil between trenches

Perimeter trench method

Figure 2.1.3   Basement excavations and timbering

retaining walls are constructed within the trench excavation and the timbering removed, the dumpling or middle can then be excavated and the base cast and joined to the retaining walls (see Fig. 2.1.3). This method could also be used in firm soils when the mechanical excavators required for bulk excavation are not available.

FOUNDATIONS

A foundation is the base on which a building rests and its purpose is to safely transfer the load of a building to a suitable subsoil.

The Building Regulations require all foundations of buildings to:

- Safely sustain and transmit to the ground the combined dead and imposed loads so as not to cause any settlement or other movement in any part of the building or of any adjoining building or works.
- Be of such a depth, or be so constructed, as to avoid damage by swelling, shrinkage or freezing of the subsoil.
- Be capable of resisting attack by deleterious material, such as sulphates, in the subsoil.

Subsoils are the soils below the topsoil; the topsoil being about 300 mm deep. Typical bearing capacities of subsoils are given in Table 2.2.1.

Table 2.2.1 Typical subsoil bearing capacities

Type	Bearing capacity (kN/m^2)
Rocks, granites and chalks	600–10,000
Non-cohesive soils; compact sands; loose uniform sands	100–600
Cohesive soils; hard clays; soft clays and silts	< 600
Peats and made ground	To be determined by investigation

Terminology

- **Backfill**   Materials excavated from site and if suitable used to fill in around the walls and foundations.
- **Bearing capacity**   Safe load per unit area which the ground can carry.
- **Bearing pressure**   The pressure produced on the ground by the loads.
- **Made ground**   Refuse, excavated rock or soil deposited for the purpose of filling in a depression or for raising the site above its natural level.

52

■ **Settlement**   Ground movement which may be caused by:
  - Deformation of the soil due to imposed loads.
  - Volume changes of the soil as a result of seasonal conditions.
  - Mass movement of the ground in unstable areas.

CHOICE OF FOUNDATION TYPE

The choice and design of foundations for domestic and small types of buildings depends mainly on two factors:

1. The total loads of the building.
2. The nature and bearing capacity of the subsoil.

The total loads of a building are taken per metre run and calculated for the worst case. The data required is:

1. Roof material dead load on the wall plus imposed load from snow, i.e. 1.5 kN/m^2 $< 30°$ pitch or $0.75 \text{ kN/m}^2 > 30°$ pitch (1 m wide strip from ridge to eaves).
2. Floor material dead load on the wall, plus an imposed loading allowance of $< 1.5 \text{ kN/m}^2$ for people and furniture (1 m wide strip from centre of the floor to the wall).
3. Wall load on the foundations (1 m wide strip of wall from top to foundation).
4. Total load on the foundations (summation of 1, 2 and 3, plus any additional allowances for wind loading that may be necessary in exposed situations).

[*Note*: Material loading due to the materials can be found in BS 648: *Weights of building materials.*]

The average total loading for a two-storey domestic dwelling of traditional construction is $30–50 \text{ kN/m}$.
  The nature and bearing capacity of the subsoil can be determined by:

■ Trial holes and subsequent investigation.
■ Bore holes and core analysis.
■ Local knowledge.

  Clay is the most difficult of all subsoils with which to deal. Down to a depth of about 1 m clays are subject to seasonal movement which occurs when the clay dries and shrinks in the summer and conversely swells in the winter with heavier rainfall. This movement occurs whenever a clay soil is exposed to the atmosphere and special foundations may be necessary.
  Subsoils which readily absorb and hold water are subject, in cold weather, to frost heave. This is a swelling of the subsoil due to the expansion of freezing water held in the soil and like the movement of clay soils it is unlikely to be even and special foundations may be needed to overcome the problem.

Ground level

Solid or
cavity wall

Depth below ground
level generally 750 to
overcome seasonal
movement and
frost heave in clay
a minimum of
1.000 will comply

P

P

P

Not less than
P and in no
case less
than 150

Width determined
by total building
load and bearing
capacity of subsoil

Mass concrete foundation
not less than 1:3:6/40 mm agg.
(15–20 N/mm²)

Strip foundations

Solid or cavity wall

D

Not less than
2 × D with a
minimum of 300

Mass concrete
foundation of
monolithic construction

Change in level

Solid or
cavity wall

P

P

P

P

P

P

Mass concrete
foundation

**Projections
and Piers**

Figure 2.2.1   Strip foundations and building regulations

TYPES OF FOUNDATION

Having ascertained the nature and bearing capacity of the subsoil the width of the foundation can be determined by either:

■ The minimum guidance as given in design tables, such as Table 12 in Approved Document A of the Building Regulations, where size of foundations are related to subsoil type, wall loading and field tests on the soil.
■ Calculating the total (dead + imposed) load per metre run of foundation and relating this to the analysed safe bearing capacity of the subsoil, i.e.:

$$\frac{\text{total load of building per metre}}{\text{safe bearing capacity of subsoil}} = \text{min. foundation width}$$

For example, if the total load is 40 kN/m and the subsoil safe bearing capacity is 80 kN/m^2, then the foundation width is:

$$\frac{40}{80} = 0.5 \text{ m} \quad \text{or} \quad 500 \text{ mm}$$

[*Note*: Safe bearing capacity is determined by obtaining the actual bearing capacity by laboratory analysis and applying a factor of safety.]

Foundations are usually made of either mass or reinforced concrete and can be considered under two headings:

■ **Shallow foundation**   Those which transfer the loads to subsoil at a point near to the ground floor of the building such as strips and rafts.
■ **Deep foundations**   Those which transfer the loads to a subsoil some distance below the ground floor of the building such as a pile.

Raft foundations are often used on poor soils for lightly loaded buildings and are considered capable of accommodating small settlements of the soil. In poor soils the upper crust of soil (450–600 mm) is often stiffer than the lower subsoil and to build a light raft on this crust is usually better than penetrating it with a strip foundation.

Typical details of the types of foundations suitable for domestic and similar buildings are shown in Figs 2.2.1–2.2.3.

Deep strip foundation

Short bored pile foundation

Figure 2.2.2 Alternative foundations for clay soils

Figure 2.2.3 Typical raft foundations

REINFORCED CONCRETE FOUNDATIONS

The function of any foundation is to transmit to the subsoil the loads of the structure. Where a structure has only light loadings such as a domestic dwelling house it is sufficient to use a mass concrete strip foundation or a simple raft. Where buildings are either heavy or transmit the loadings at a series of points such as in a framed building it is uneconomic to use mass concrete. The plan size of a foundation is a constant feature being derived from:

$$\frac{\text{point or column load}}{\text{safe bearing capacity of subsoil}}$$

For example, if the column shown in Fig. 2.3.1 transmits a 50 kN load to subsoil of safe bearing capacity 80 kN/m^2, then the square column foundation dimensions are:

$$\sqrt{\frac{50}{80}} = 0.79 \text{ m} \quad \text{or} \quad 790 \text{ mm (800 mm square)}$$

The thickness of a mass concrete foundation for a heavy point load would result in a foundation which is costly and adds unnecessary load to the subsoil. Reinforced concrete foundations are generally cheaper and easier to construct than equivalent mass concrete foundations but will generally require the services of a structural engineer to formulate an economic design. The engineer must define the areas in which tension occurs and specify the reinforcement required, since concrete is a material which is weak in tension. Further studies of reinforced concrete principles can be found in Chapter 9.2.

TYPES OF FOUNDATION

The principal types of reinforced concrete foundation for buildings are:

1. Strip foundations.
2. Isolated or pad foundations.
3. Raft foundations.

Heavy wall

Reinforced concrete
(1:2:4/20 mm agg.
– 25 N/mm²)
strip foundation

Weak
concrete
blinding

Distribution bars

Main bars

RC strip foundation

75 mm kicker

Starter bars
for column

Ground floor
position

Reinforced concrete
(1:2:4/20 mm agg.
– 25 N/mm²)
column foundation

Reinforced
concrete column

Weak concrete
blinding

Main bars in
both directions

RC isolated or pad foundation

Figure 2.3.1 Reinforced concrete strip and pad foundations

4. Combinations of 1, 2 and 3.
5. Piled foundations.

The foundations listed in 4 and 5 above are included in *Advanced Construction Technology* and are therefore not considered in this volume.

STRIP FOUNDATIONS

Reinforced concrete strip foundations are used to support and transmit the loads from heavy walls. The effect of the wall on the relatively thin foundation is to act as a point load and the resultant ground pressure will induce tension on the underside across the width of the strip. Tensile reinforcement is therefore required in the lower face of the strip with distribution bars in the second layer running longitudinally (see Fig. 2.3.1). The reinforcement will also assist the strip in spanning any weak pockets of soil encountered in the excavations.

ISOLATED OR PAD FOUNDATIONS

This type of foundation is used to support and transmit the loads from piers and columns. The most economic plan shape is a square but if the columns are close to the site boundary it may be necessary to use a rectangular plan shape of equivalent area. The reaction of the foundation to the load and ground pressures is to cup, similar to a saucer, and therefore main steel is required in both directions. The depth of the base will be governed by the anticipated moments and shear forces, the calculations involved being beyond the scope of this volume. Incorporated in the base will also be the starter bars for a reinforced concrete column or the holding down bolts for a structural steel column (see Fig. 2.3.1).

RAFT FOUNDATIONS

The principle of any raft foundation is to spread the load over the entire area of the site. This method is particularly useful where the column loads are heavy and thus requiring large bases or where the bearing capacity is low, again resulting in the need for large bases. Raft foundations can be considered under three headings:

1. Solid slab rafts.
2. Beam and slab rafts.
3. Cellular rafts.

Solid slab rafts are constructed of uniform thickness over the whole raft area, which can be wasteful since the design must be based on the situation existing where the heaviest load occurs. The effect of the load from columns and the ground pressure is to create areas of tension under the columns and areas of tension in the upper part of the raft between the columns. Very often a nominal mesh of reinforcement is provided in the faces where tension does not occur to control shrinkage cracking of the concrete (see Fig. 2.3.2).

Beam and slab rafts are an alternative to the solid slab raft and are used where poor soils are encountered. The beams are used to distribute the column loads over the area of the raft, which usually results in a reduction of the slab thickness. The beams can be upstand or downstand depending upon the bearing capacity of the soil near the surface. Downstand beams will give a saving on excavation costs whereas upstand beams create a usable void below the ground floor if a suspended slab is used (see Fig. 2.3.2).

RC solid slab raft foundation

Column starter bars

Reinforced concrete (1:2:4/20 mm agg. – 25 N/mm²) solid slab raft

75 mm high kicker

Weak concrete blinding

Column starter bars

RC suspended slab of *in-situ* or precast concrete

Columns positioned on beam intersection

RC upstand beams in 2 directions

75 mm high kicker

Weak concrete blinding

Reinforced concrete raft (1:2:4/20 mm agg. – 25 N/mm²)

RC beam and slab raft foundation

Figure 2.3.2   Reinforced concrete raft foundations

CELLULAR RAFTS

This form of foundation can be used where a reasonable bearing capacity subsoil can only be found at depths where beam and slab techniques become uneconomic. The construction is similar to reinforced concrete basements except

Figure 2.3.3 Typical cellular raft details

that internal walls are used to spread the load over the raft and divide the
void into cells. Openings can be formed in the cell walls allowing the voids to
be utilized for the housing of services, store rooms or general accommodation
(see Fig. 2.3.3).

BLINDING

A blinding layer 50 to 75 mm thick of weak concrete or coarse sand should be placed under all reinforced concrete foundations. The functions of the blinding are to fill in any weak pockets encountered during excavations and to provide a true level surface from which the reinforcement can be positioned. If formwork is required for the foundation some contractors prefer to lay the blinding before assembling the formwork; the alternative is to place the blinding within the formwork and allow this to set before positioning the reinforcement and placing the concrete.

CONCRETE

Concrete is a mixture of cement, fine aggregate, coarse aggregate and water. The proportions of each material control the strength and quality of the resultant concrete.

PORTLAND CEMENT

Cement is the setting agent of concrete and the bulk of cement used in this country is Portland cement. This is made from chalk or limestone and clay and is generally produced by the wet process.

In this process the two raw materials are washed, broken up and mixed with water to form a slurry. This slurry is then pumped into a steel rotary kiln which is from 3 to 4 m in diameter and up to 150 m long and lined with refractory bricks. While the slurry is fed into the top end of the kiln a pulverised coal is blown in at the bottom end and fired. This raises the temperature at the lower end of the kiln to about 1,400 °C. The slurry passing down the kiln first gives up its moisture, then the chalk or limestone is broken down into carbon dioxide and lime, and finally forms a white hot clinker which is transferred to a cooler before being ground. The grinding is carried out in a ball mill which is a cylinder some 15 m long and up to 4.5 m in diameter containing a large number of steel balls of various sizes, which grind the clinker into a fine powder. As the clinker is being fed into the ball mill gypsum (about 5%) is added to prevent a flash setting off the cement.

The alternative method for the preparation of Portland cement is the dry process. The main difference between this and the wet process is the reduction in the amount of water that has to be driven off in the kiln. A mixture of limestone and shale is used which is proportioned, ground and blended to form a raw meal of low moisture content. The meal is granulated in rotating pans with a small amount of water before being passed to a grate for preheating prior to entering the kiln. The kiln is smaller than that used in the wet process but its function is the same, that is, to form a clinker, which is then cooled, ground and mixed with a little gypsum as described for the previous process.

Rapid-hardening Portland cement is more finely ground than ordinary Portland cement. Its main advantage is that it gains its working strength earlier than ordinary cement. The requirements for both ordinary Portland and rapid-hardening Portland cement are given in BS EN 197-1.

HIGH-ALUMINA CEMENT – BS 915

This is made by firing limestone and bauxite (aluminium ore) to a molten state, casting it into pigs and finally grinding it into a fine powder. Its rate of hardening is very rapid and produces a concrete which is resistant to the natural sulphates found in some subsoils. It can, however, cost up to two and a half times as much as ordinary Portland cement. However, high-alumina cement has its limitations. When hydrated, the cement converts or undergoes a chemical change which can reduce the concrete's strength and resistance to aggressive chemicals. The manner and extent of conversion varies with the amount of free water/cement ratio, the temperature in which the concrete is located and the amount of moisture or humidity in the surrounding environment. All these factors are applicable during manufacture, subsequent curing and its design life. Owing to a number of structural failures, notably in the humid atmospheres of swimming pools, high-alumina cement is now not normally permitted in exposed concrete structures.

Other forms of cement available are as follows:

- Portland blastfurnace – BS 146.
- Low-heat Portland – BS 1370.
- Low-alkali sulphate-resisting – BS 4027.
- High-slag blastfurnace – BS 4246.
- Supersulphated – BS 4248.
- Masonry – BS 5224.
- Portland pulverized fuel ash – BS 6588.
- Pozzolanic pulverized fuel ash – BS 6610.
- Portland limestone – BS 7583.

Cement should be stored on a damp-proof floor in the dry and kept for short periods only because eventually cement will harden as a result of the action of moisture in the air. This is known as air hardening and any hardened cement should be discarded.

AGGREGATES

These are the materials which are mixed with the cement to form concrete and are classed as a fine or coarse aggregate. Fine aggregates are those which will pass a standard 5 mm sieve and coarse aggregates are those which are retained on a standard 5 mm sieve. All-in aggregate is a material composed of both fine and coarse aggregates.

A wide variety of materials (for example, gravel, crushed stone, brick, furnace slag and lightweight substances, such as foamed slag, expanded clay and vermiculite) are available as aggregates for the making of concrete.

In making concrete aggregates must be graded so that the smaller particles of the fine aggregate fill the voids created by the coarse aggregate. The cement paste fills the voids in the fine aggregate thus forming a dense mix.

Aggregates from natural sources are covered in BS 882, with a variety of lightweight aggregates detailed in BS 3797. Air-cooled blastfurnace slag aggregate of normal weight is specified in BS 1047.

WATER

The water used in the making of concrete must be clean and free from impurities which could affect the concrete. It is usually specified as being of a quality fit for drinking. A proportion of the water will set up a chemical reaction which will harden the cement. The remainder is required to give the mix workability and will evaporate from the mix while it is curing, leaving minute voids. An excess of water will give a porous concrete of reduced durability and strength.

The quantity of water to be used in the mix is usually expressed in terms of the water/cement ratio which is:

$$\frac{\text{the total weight of water in the concrete}}{\text{weight of cement}}$$

for most mixes the ratio is between 0.4 and 0.7.

Concrete mixes can be expressed as volume ratios thus:

- 1:2:4 = 1 part cement, 2 parts fine aggregate and 4 parts coarse aggregate.
- 1:5   = 1 part cement and 5 parts all-in aggregate.

Some common mixes

- 1:10 – not a strong mix but it is suitable for filling weak pockets in excavations and for blinding layers.
- 1:8 – slightly better than the last, suitable for paths and pavings.
- 1:6 – a strong mix suitable for mass concrete foundations, paths and pavings.
- 1:3:6 – the weakest mix equivalent to that quoted in Approved Document A as deemed to satisfy the requirements of Building Regulation A1.
- 1:2:4 – strong mix which is practically impervious to water, in common use especially for reinforced concrete.

Mixing concrete

Concrete can be mixed or batched by two methods:

1. By volume.
2. By mass.

A 25-kg bag of cement has a volume of approximately 0.02 m^3.

Batching by volume

This method is usually carried out using an open bottom box (of such dimensions as to make manual handling possible) called a 'gauge box'. For a 1:2:4 mix a gauge box is filled once with cement, twice with fine aggregate and four times with coarse aggregate, the top of the gauge box being struck off level each time.

If the fine aggregate is damp or wet its volume will increase by up to 25% and therefore the amount of fine aggregate should be increased by this amount. This increase in volume is called bulking.

Batching by mass

This method involves the use of a balance which is linked to a dial giving the exact mass of the materials as they are placed in the scales. This is the best method since it has a greater accuracy and the balance can be attached to the mixing machine.

Hand mixing

This should be carried out on a clean hard surface. The materials should be thoroughly mixed in the dry state before the water is added. The water should be added slowly, preferably using a rose head until a uniform colour is obtained.

Machine mixing

The mix should be turned over in the mixer for at least two minutes after adding the water. The first batch from the mixer tends to be harsh since some of the mix will adhere to the sides of the drum. This batch should be used for some less important work such as filling in weak pockets in the bottom of the excavation.

Ready mixed

This is used for large batches with lorry transporters up to 6 m^3 capacity. It has the advantage of eliminating site storage of materials and mixing plant, with the guarantee of concrete manufactured to quality-controlled standards. Placement is usually direct from the lorry, therefore site-handling facilities must be co-ordinated with deliveries.

Handling

If concrete is to be transported for some distance over rough ground the runs should be kept as short as possible since vibrations of this nature can cause segregation of the materials in the mix. For the same reason concrete should not be dropped from a height of more than 1 m. If this is unavoidable a chute should be used.

Placing

If the concrete is to be placed in a foundation trench it will be levelled from peg to peg (see Fig. 1.4.2) or if it is to be used as an oversite bed the external walls could act as a levelling guide. The levelling is carried out by tamping with a straight edge board; this tamping serves the dual purpose of compacting and bringing the excess water to the surface so that it can evaporate. Concrete must not be over-tamped as this will not only bring the water to the surface but also the cement paste which is required to act as the matrix. Concrete should be placed as soon as possible after mixing to ensure that the setting action has not commenced. Concrete which dries out too quickly will not develop its full strength, therefore new concrete should be protected from the drying winds and sun by being covered with canvas, straw, polythene sheeting or damp sawdust. This protection should be continued for at least three days since concrete takes about twenty-eight days to obtain its working strength.

Specifying concrete

Any of four methods may be used to specify concrete:

- Designed mix
- Prescribed mix
- Standard mix
- Designated mix

Designed mix

The mix is specified by a grade corresponding to required characteristic compressive strength at 28 days. There are 12 grades from C7.5 to C60, the C indicating compressive and the number the strength in N/mm^2 or MPa. Flexural (F) strength grades may also be specified as F3, F4 or F5, i.e. 3, 4 or 5 N/mm^2. In addition to stating the strength grades the purchaser must also specify any particular requirements for cement and aggregate content and maximum free water/cement ratio.

Prescribed mix

This is a recipe of constituents with their properties and quantities used to manufacture the concrete. The specifier must state:

- The type of cement.
- Type of aggregates and their maximum size.
- Mix proportions by weight.
- Degree of workability (slump and/or water cement ratio) and the application.

Prescribed mixes are based on established data indicating conformity to strength, durability and other characteristics. Basic examples could include 1:3:6/40 mm aggregate and 1:2:4/20 mm aggregate.

Standard mix

Mixes are produced from one of five grades, ranging from ST1 to ST5, with corresponding 28 day strength characteristics of 7.5 to a limit of only 25 N/mm^2. Mix composition and details are specified by:

- Cement to aggregate by weight.
- Type of cement.
- Aggregate type and maximum size.
- Workability.
- Use or omission of reinforcement.

These mixes are most suited to site production, where the scale of operations is relatively small. Alternatively, they may be used where mix design procedures would be too time consuming, inappropriate or uneconomic.

Designated mix

These mixes are selected relative to particular applications and site conditions, in place of generalizations or use of alternative design criteria that may not be entirely appropriate. Grading and strength characteristics are extensive and vary with application as indicated:

- General (GEN), graded 0–4, ranging from 7.5 to 25 N/mm^2 characteristic strength. For foundations, floors and external works.
- Foundations (FND), graded 2, 3, 4A and 4B with characteristic strength of 35 N/mm^2. These are particularly appropriate for resisting the effects of sulphates in the ground.
- Paving (PAV), graded 1 or 2 in 35 or 45 N/mm^2 strengths, respectively. A strong concrete for use in driveways and heavy duty pavings.
- Reinforced (RC) and prestressed concrete graded 30, 35, 40, 45 and 50 corresponding with characteristic strength and exposures ranging from mild to most severe.

In addition to application, the purchaser needs to specify:

- Reinforced or unreinforced.
- Prestressed.
- Heated or ambient temperature.
- Maximum aggregate size (if not 20 mm).
- Exposure to chemicals (chlorides and sulphates).
- Exposure to subzero temperatures (placing and curing).

Due to the precise nature of these designated mix concretes, quality control is of paramount importance. Therefore, producers are required to have quality assurance product conformity accreditation to BS EN ISO 9001.

[*Note*: Further details of concrete grades and analysis are found in BS 5328: Parts 1 and 2.]

RETAINING WALLS

The basic function of a retaining wall is to retain soil at a slope which is greater than it would naturally assume, usually at a vertical or near vertical position. The natural slope taken up by any soil is called its angle of repose and is measured in relationship to the horizontal. Angles of repose for different soils range from 45° to near 0° for wet clays but for most soils an average angle of 30° is usually taken. It is the wedge of soil resting on this upper plane of the angle of repose which a retaining wall has to support. The walls are designed to offer the necessary resistance by using their own mass to resist the thrust or relying upon the principles of leverage. The terminology used in retaining wall construction is shown in Fig. 2.5.1.

DESIGN PRINCIPLES

The design of any retaining wall is basically concerned with the lateral pressures of the retained soil and any subsoil water. It must be designed to ensure that:

- Overturning does not occur.
- Sliding does not occur.
- The soil on which the wall rests is not overloaded.
- The materials used in construction are not overstressed.

It is difficult to accurately define the properties of any soil since they are variable materials and the calculation of pressure exerted at any point on the wall is a task for the expert who must take into account the following factors:

- Nature and type of soil.
- Height of water table.
- Subsoil water movements.
- Type of wall.
- Materials used in the construction of the wall.

Design calculations relate to the resultant thrust of retained material behind a one metre length of wall. This can be calculated by applying Rankine's long–established formula:

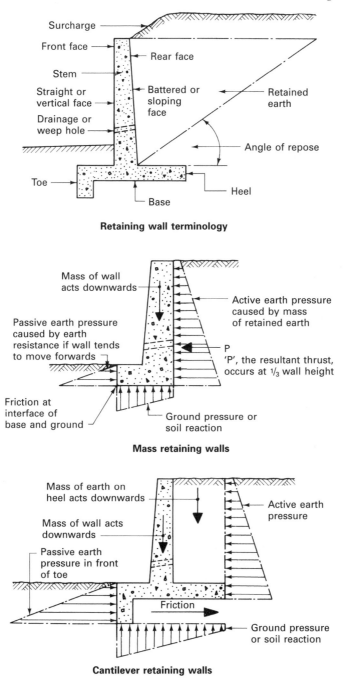

Figure 2.5.1 Retaining wall terminology and pressures

$$P = \frac{wh^2}{2} \times \frac{1 - \sin \theta}{1 + \sin \theta}$$

where P = resultant thrust
   w = density of retained soil
   h = height of wall
   θ = soil angle of repose

For example, if $w = 1,400$ kg/m^3, $h = 2$ m and $\theta = 30°$, then:

$$P = \frac{1,400 \times (2)^2}{2} \times \frac{1 - 0.5}{1 + 0.5} = 934 \text{ kg}$$

kg $\times$ gravity = Newtons, therefore

   $934 \times 9.81 = 9.16$ kN

Structural design is based on the overturning or bending moment (BM) which occurs at one-third wall height, if the earth pressure is uniform and the diagram triangular, as shown in Fig. 2.5.1.
   In this example, $\frac{1}{3} \times 2$ m $= 0.67$ m. Therefore,

   BM $= 9.16$ kN $\times 0.67$ m $= 6.14$ kN m

P, the resultant thrust for water, can be calculated from the simple formula:

$$P = \frac{wh^2}{2}$$

where w = density of water (1,000 kg/m^3).

EARTH PRESSURES

The designer is mainly concerned with the effect of two forms of earth pressure:

1. Active earth pressure.
2. Passive earth pressure.

   Active earth pressures are those which at all times are tending to move or overturn the retaining wall and are composed of the earth wedge being retained together with any hydrostatic pressure caused by the presence of ground water. The latter can be reduced by the use of subsoil drainage behind the wall or by inserting drainage openings called weep holes through the thickness of the stem enabling the water to drain away.
   Passive earth pressures are reactionary pressures which will react in the form of a resistance to movement of the wall. If the wall tends to move forward, the earth in front of the toe will be compressed and a reaction in the form of passive pressure will build up in front of the toe to counteract the forward movement. This pressure can be increased by enlarging the depth of the toe or by forming a rib on the underside of the base. Typical examples of these pressures are shown in Fig. 2.5.1.

STABILITY

The overall stability of a retaining wall is governed by the result of the action and reaction of a number of loads:

Applied loads   For example, soil and water pressure on the back of the wall; the mass of the wall and in the case of certain forms of cantilever walls the mass of the soil acting with the mass of the wall.

Induced loads   For example, the ground pressure under the base, the passive pressure at the toe and the friction between the underside of the base and the soil.

EFFECTS OF WATER

Ground water behind a retaining wall, whether static or percolating through a subsoil, can have adverse effects upon the design and stability. It will increase the pressure on the back of the wall and by reducing the soil shear strength it can reduce the bearing capacity of the soil; it can reduce the frictional resistance between the base and the soil and reduce the possible passive pressure in front of the wall. It follows therefore that the question of drainage of the water behind the retaining wall is of the utmost importance in the design.

SLIP CIRCLE FAILURE

This is a form of failure sometimes encountered with retaining walls in clay soils particularly if there is a heavy surcharge. It takes the form of a rotational movement of the soil and wall along a circular arc which starts behind the wall and passes under the base, resulting in a tilting and forward movement of the wall. If the movement is not unacceptable further movement can be prevented by driving sheet piles into the ground in front of the toe to a depth that will cut the slip circle arc.

TYPES OF WALL

MASS RETAINING WALLS

These are sometimes called gravity walls and rely upon their own mass together with the friction on the underside of the base to overcome the tendency to slide or overturn. They are generally only economic up to a height of 1.800 m. Mass walls can be constructed of semi-engineering quality bricks bedded in a 1:3 cement mortar or of mass concrete. The latter could have some light fabric reinforcement to control surface cracking. Natural stone is suitable for small walls up to 1.000 m high but generally it is used as a facing material for walls over 1.000 m. Typical examples of this are shown in Fig. 2.5.2.

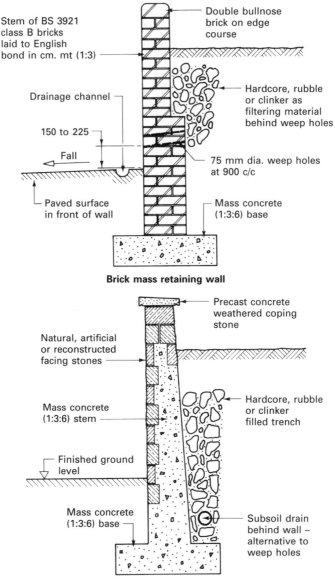

Stem of BS 3921
class B bricks
laid to English
bond in cm. mt (1:3)

Double bullnose
brick on edge
course

Drainage channel

Hardcore, rubble
or clinker as
filtering material
behind weep holes

150 to 225

Fall

75 mm dia. weep holes
at 900 c/c

Paved surface
in front of wall

Mass concrete
(1:3:6) base

Brick mass retaining wall

Precast concrete
weathered coping
stone

Natural, artificial
or reconstructed
facing stones

Mass concrete
(1:3:6) stem

Hardcore, rubble
or clinker
filled trench

Finished ground
level

Mass concrete
(1:3:6) base

Subsoil drain
behind wall –
alternative to
weep holes

Mass concrete retaining wall with stone facings

Figure 2.5.2   Typical mass retaining walls

CANTILEVER WALLS

These are usually of reinforced concrete and work on the principles of leverage.
Two basic forms can be considered, a base with a large heel so that the mass
of earth above can be added to the mass of the wall for design purposes, or, if

Figure 2.5.3 Typical reinforced concrete cantilever retaining walls

this form is not practicable, a cantilever wall with a large toe must be used (see Fig. 2.5.3). The drawings show typical sections and patterns of reinforcement encountered with these basic forms of cantilever retaining walls. The main steel occurs on the tension face of the wall and nominal steel (0.15% of the cross-sectional area of the wall) is very often included in the opposite face to control the shrinkage cracking which occurs in *in-situ* concrete work. Reinforcement requirements, bending, fabricating and placing are dealt with in detail in the section on reinforced concrete.

Reinforced cantilever walls have an economic height range of 1.200 to 6.000 m; walls in excess of this height have been economically constructed using prestressing techniques. Any durable facing material may be applied to the surface to improve the appearance of the wall but it must be remembered that such finishes are decorative and add nothing to the structural strength of the wall.

COUNTERFORT RETAINING WALLS

These walls can be constructed of reinforced or prestressed concrete and are considered suitable if the height is over 4.500 m. The counterforts are triangular beams placed at suitable centres behind the stem and above the base to enable the stem and base to act as slabs spanning horizontally over or under the counterforts. Figure 2.5.4 shows a typical section and pattern of reinforcement for a counterfort retaining wall.

If the counterforts are placed on the face of the stem they are termed buttresses and the whole arrangement is called a buttress-retaining wall. In both formats the design and construction principles are similar.

PRECAST CONCRETE RETAINING WALLS

These are manufactured from high-grade precast concrete on the cantilever principle usually to a 600 mm wide module (see Fig. 2.5.5). They can be erected on a foundation as a permanent retaining wall or be free standing to act as a dividing wall between heaped materials such as aggregates for concrete. In the latter situation they can increase by approximately three times the storage volume for any given area. Other advantages are a reduction in time by eliminating the curing period which is required for *in-situ* walls and eliminating the need for costly formwork together with the time required to erect and dismantle the temporary forms. The units are reinforced on both faces to meet all forms of stem loading. Lifting holes are provided which can be utilized as strap-fixing holes if required. Special units to form internal angles, external angles, junctions and curved walls are also available to provide flexible layout arrangements.

PRECAST CONCRETE CRIB-RETAINING WALLS

Crib walls are designed on the principle of a mass retaining wall. They consist of a framework or crib of precast concrete or timber units within which the soil

Figure 2.5.4 Typical reinforced concrete counterfort retaining wall

is retained. They are constructed with a face batter of between 1:6 and 1:8 unless the height is less than the width of the crib ties when the face can be constructed vertical. Subsoil drainage is not required since the open face provides for adequate drainage (see Fig. 2.5.5).

Typical 'Marley' precast concrete retaining wall

Note: All units connected with dowels

'Anda-Crib' precast concrete retaining wall

Figure 2.5.5   Precast concrete retaining walls

CHAPTER 2.6

BASEMENTS

Appendix E in Approved Document B of the Building Regulations defines a basement storey as a storey with a floor which at some point is more than 1.200 m below the highest level of ground adjacent to the outside walls. This definition is given in the context of inhibiting the spread of fire within a building and generally the fire resistance requirements for basements are more onerous than the ground or upper storeys in the same building. This section on basements is only concerned with basement storeys which are below ground level.

The structural walls of a basement below ground level are in fact retaining walls which have to offer resistance to the soil and ground water pressures as well as assisting to transmit the superstructure loads to the foundations. It is possible to construct a basement free of superstructural loadings but these techniques are beyond the scope of this book.

WATERPROOFING

Apart from the structural design of the basement walls and floor, waterproofing presents the greatest problem in basement construction. Building Regulation C2 requires such walls to be constructed so that they will not transmit moisture from the ground to the inside of the building or to any material used in the construction that would adversely be affected by moisture. Building Regulation C2 also imposes similar conditions on the construction of floors. Basement structures can be waterproofed by one of three basic methods:

- Monolithic structures.
- Drained cavities.
- Membranes.

MONOLITHIC STRUCTURES

These are basements of dense reinforced concrete using impervious aggregates for the walls and floor to form the barrier to water penetration. Great care must be taken with the design of the mix, the actual mixing and placing together with

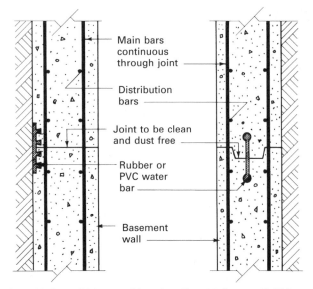

Note: Horizontal joints positioned at 12 to 15 times wall thickness

Construction joints

Shrinkage joint

Figure 2.6.1 Joints and water bars

careful selection and construction of the formwork if a satisfactory water barrier is to be achieved. Shrinkage cracking can largely be controlled by forming construction joints at regular intervals. These joints should provide continuity of reinforcement and by the incorporation of a PVC or rubber water bar a barrier to the passage of water; typical examples are shown in Fig. 2.6.1. Monolithic structures, while providing an adequate barrier to the passage of water, are not always vapourproof.

DRAINED CAVITIES

This method provides an excellent barrier to moisture penetration of basements by allowing any moisture which has passed through the structural wall to drain down within a cavity formed between the inner face of the structural wall and an inner non-loadbearing wall. This internal wall is built of a floor covering of special triangular precast concrete tiles which allows the moisture, from the cavity, to flow away under the tiles to a sump where it is discharged into a drainage system either by gravity or pumping. This method of waterproofing is usually studied in detail during advanced courses in construction technology.

MEMBRANES

A membrane is a relatively thin material placed either on the external or internal face of a basement wall or floor to provide the resistance to the passage of moisture to the inside of the basement. If the membrane is applied externally protection is also given to the structural elements and the hydrostatic pressure will keep it firmly in place, but a reasonable working space must be allowed around the perimeter of the basement. This working space will entail extra excavation and subsequent backfilling after the membrane has been applied. If adequate protection is not given to the membrane it can easily be damaged during the backfilling operation. An internally applied membrane gives no protection to the structural elements and there is the danger that the membrane may be forced away from the surfaces, by water pressure, unless it is adequately loaded. These loading coats will reduce the usable volume within the basement (see Figs 2.6.2 and 2.6.3).

Suitable materials which can be used for forming membranes are fibre-reinforced bituminous felt, polythene sheet, polyisobutylene plastic, epoxy resin compounds, bituminous compounds and mastic asphalt. Also, more recent developments for impervious membrane applications include polymer-modified bitumen with polyester reinforcement. This is available as styrene–butadiene–styrene (SBS) or atactic polypropylene (APP).

ASPHALT TANKING

Asphalt is a natural or manufactured mixture of bitumen with a substantial proportion of inert mineral matter. When heated, asphalt becomes plastic and can be moulded by hand pressure into any shape. Bitumen is a complex mixture of hydrocarbons and has both waterproofing and adhesive properties. In its natural state asphalt occurs as a limestone rock impregnated with bitumen and is mined notably in France, Switzerland and Sicily. Another source of asphalt is the asphalt lake in Trinidad in the West Indies which was discovered by Sir Walter Raleigh in 1595 and today still yields about 100,000 tonnes annually. In the centre of this lake the asphalt is a liquid but nearer the edges it is a semi-fluid, and although large quantities are removed during the day the lake refills during the night. The lake asphalt is refined or purified in Trinidad and shipped in barrels for use in building

Figure 2.6.2   External tanking in mastic asphalt

construction all over the world. Natural rock asphalt is crushed and processed to remove unwanted mineral matter before being compounded into mastic asphalt. Bitumen for use with mastic asphalt is also made on a large scale as a residue in the distillation of petroleum.

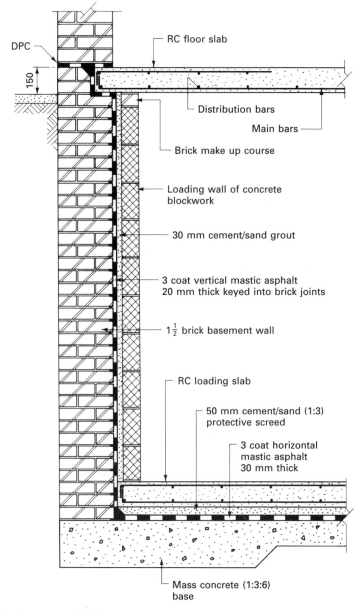

Figure 2.6.3   Internal tanking in mastic asphalt

Mastic asphalt is a type of asphalt composed of suitably graded mineral matter and asphaltic cement to form a coherent, voidless and impermeable mass. The asphalt cement consists of bitumen, lake asphalt, asphaltite or blends of these, sometimes with the addition of flux oil, which is used for softening bitumen or rendering it less viscous. Fine aggregates of natural rock asphalt and limestone combine with coarse aggregates of igneous and calcareous angular stones if required. Naturally occurring graded siliceous material can also be added.

The basic principle of asphalt tanking is to provide a continuous waterproof membrane to the base and walls of the basement. Continuity between the vertical and horizontal membranes is of the utmost importance, and since asphalt will set rapidly once removed from the heat source used to melt the blocks, it is applied in layers over small areas; again continuity is the key factor to a successful operation. Joints in successive coats should be staggered by at least 150 mm in horizontal work and at least 75 mm in vertical work.

On horizontal and surfaces up to 30° from the horizontal three coats of asphalt should be applied to give a minimum total thickness of 30 mm. Vertical work should also be a three-coat application to give a total thickness of 20 mm. The junction between horizontal and vertical work should be strengthened by a two-coat angle fillet forming a 50×50 mm chamfer. To prevent curling and consequent infiltration of moisture behind the vertical tanking the top edge should be turned into a splayed chase or groove 25 mm wide $\times$ 25 mm deep.

It is essential that vertical asphalt is suitably keyed to its background. Concrete formed by using sawn boards for the formwork will usually provide an acceptable surface but smooth concrete will need treatment such as bush hammering the surface and washing to remove all loose particles. Alternatively a primer of sand/cement plastic emulsion or pitch/polymer rubber emulsion can be used. Brick walls can be constructed of keyed bricks or the joints can be raked out to a depth of 20 mm as the work proceeds to provide the necessary keyed surface.

During the construction period the asphalt tanking must be protected against damage from impact, following trades and the adverse effects of petrol and oil. Horizontal asphalt tanking coats should be covered with a fine concrete screed at least 50 mm thick as soon as practicable after laying. Vertical asphalt tanking coats should be protected by building a half brick or block wall 30 mm clear of the asphalt; the cavity so formed should be filled with a mortar grout as the work proceeds to ensure perfect interface contact. In the case of internal tanking this protective wall will also act as the loading coat.

Any openings for the passages of pipes or ducts may allow moisture to penetrate unless adequate precautions are taken. The pipe or duct should be primed and coated with three coats of asphalt so that the sleeve formed extends at least 75 mm on either side of the tanking membrane before being placed in the wall or floor. The pipe or duct is connected to the tanking by a two-coat angle fillet (see Fig. 2.6.2).

The main advantages of mastic asphalt as a waterproof membrane are:

1. It is a thermoplastic material and can therefore be heated and reheated if necessary to make it pliable for moulding with a hand float to any desired shape or contour.
2. Durability: bituminous materials have been used in the construction of buildings for over 5,000 years and have remained intact to this day as shown by excavations in Babylonia.
3. Impervious to both water and water vapour.

4. Non-toxic, vermin and rot proof and it is odourless after laying.
5. It is unaffected by sulphates in the soil which, if placed externally, will greatly improve the durability of a concrete structure.

The application of mastic asphalt is recognized as a specialist trade in the building industry and therefore most asphalt work is placed in the hands of specialist subcontractors, most of which are members of the Mastic Asphalt Council and Employers Federation Limited. The Federation is a non-profit-making organization whose objectives are to provide technical information and promote the use of mastic asphalt as a high-quality building material.

OTHER SHEET MEMBRANES

Plastic and bitumen sheeting materials are suited to shallow basements. The base structure of concrete or masonry is prepared with a primer of bituminous solution before sheeting and is hot bitumen bonded with 100 mm side and 150 mm end lapping in at least two layers. Figure 2.6.4 shows application to wall and floor.

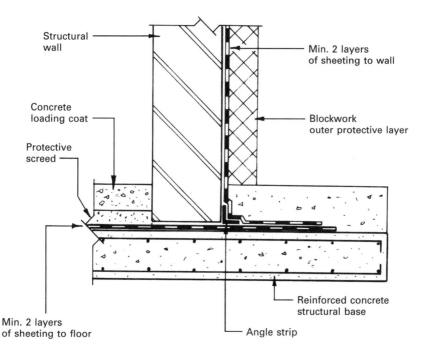

Figure 2.6.4 Impervious membrane tanking to shallow basement

TREES: EFFECT ON FOUNDATIONS

Close proximity of trees to buildings can result in direct physical damage to foundations, drains, paths, drives, etc., by growth and expansion of root systems. With coventional strip foundations and those provided for pre-1950s dwellings, where the concrete was rarely placed deeper than 0.5 m, it is advisable to maintain a distance at least equal to the mature tree's height between building and tree. Some relaxation may be applied where tree species can be identified as having less water demand. Guidance is given in Table 2.7.1 with approximate heights of mature trees.

Table 2.7.1 Tree characteristics

Species	Approximate mature height (m)	Water demand: H = high, M = medium, L = low
Elm	30	H
Oak	30	H
Poplar	30	H
Ash	30	M
Cedar	30	M
Chestnut	30	M
Hemlock	30	M
Lime	30	M
Larch	30	L
Scots Pine	30	L
Cypress	25	H
Willow	25	H
Maple	25	M
Sycamore	25	M
Walnut	25	M
Beech	25	L
Birch	25	L
Holly	20	L
Cherry	15	M
Hawthorn	15	M
Rowan	15	M

Note: Where groups of trees occur and the subsoil is known to be moisture responsive (i.e. shrinkable clay), the suggested building distance is 1.5 × mature height.

Figure 2.7.1   Area of UK with zones of shrinkable clay subsoils

In the UK, shrinkable clays occur in specific areas found roughly below an imaginary line linking the Humber estuary with a point intersecting the mouth of the River Severn, as shown in Fig. 2.7.1.

Clay subsoils are particularly prone to changes in moisture content, with a high degree of shrinkage apparent during periods of drought. When added to the effect of moisture abstraction by vegetation (trees), voids and swallow holes are to be expected. Clay heave is the opposite phenomenon, partly attributed to saturation from inclement weather. It is more likely caused by removal of trees within the immediate 'footprint' of new buildings or the immediate surrounds. Figure 2.7.2 shows cracking patterns in buildings due to these indirect sources of damage.

The accepted depth for foundations in clay subsoils is at least 900 mm in the absence of trees. However, where trees are apparent or have been within the preceding three years, the minimum depth of foundation must be calculated. Where insurers issue structural guarantees, this depth can be determined from tables provided in procedure manuals issued by the National House Building Council or Zurich Municipal. Figure 2.7.3 provides some guidance, as indication

Subsoil dehydration and shrinkage

Ground heave

Figure 2.7.2 Indirect damage to buildings

Species of tree	Depth of foundation trench (m)						
	d/h = 0.1	0.25	0.33	0.5	0.66	0.75	1.0
Poplar, elm willow and oak	Not acceptable	2.8	2.6	2.3	2.1	1.9	1.5
Others	Not acceptable	2.4	2.1	1.8	1.5	1.2	1.0

Figure 2.7.3   Tree proximity and foundation depth

only, of the likely depths of foundation excavation necessary where tree distance (*D*) to height (*H*) ratios are less than unity. Beyond 1.5 m depth, conventional strip foundations will not be economical or practically viable, but trench-filled concrete will be appropriate. For foundations over 2 m, short-bored piles with ground beams are preferred.

HEAVE PRECAUTIONS

In addition to foundation depths exceeding soil movement zones, the areas closer to the surface must be provided with additional treatment. Here, prior to concreting, the trench is prepared with a lining of low-density expanded polystyrene (or other acceptable compressible material) along with an outer polythene sheet sleeving. The polystyrene is designed to absorb seasonal clay movement and the polythene acts as a slip membrane preventing soil adhesion to the foundation. Figure 2.7.4 shows application to trench fill and piled foundations.

Precast concrete
suspended floor
(see Ch. 4.3)

External
cavity wall

DPC

150 mm
min. void

Compressible
material on
inside face of
foundation

Polythene sheet
prevents soil
adhesion to concrete

Mass concrete (reinforced
if required) foundation

Note: Compressible material may be required to both sides of foundation

Deep strip or trench fill

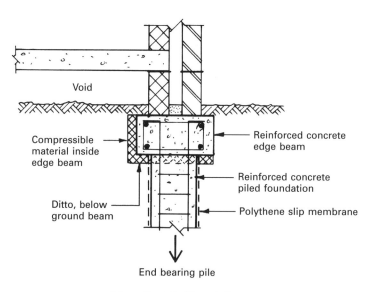

Void

Compressible
material inside
edge beam

Reinforced concrete
edge beam

Reinforced concrete
piled foundation

Ditto, below
ground beam

Polythene slip membrane

End bearing pile

Alternative piled foundation

Figure 2.7.4   Precautionary treatment to foundations in shrinkable clay

PART 3

SUPERSTRUCTURE

STONEWORK, BRICKWORK AND BLOCKWORK

STONEWORK

The natural formation of stones or rocks is a very long process which commenced when the earth was composed only of gases. These gases eventually began to liquefy forming the land and the sea, the land being only a thin crust or mantle to the still molten core.

Igneous rocks originated from the molten state; plutonic rocks are those in the lower part of the earth's mantle; whereas hypabyssal rocks solidified rapidly near the upper surface of the crust.

BUILDING STONES

Stones used in building can be divided into three classes as follows:

- Igneous.
- Sedimentary.
- Metamorphic.

Igneous stones

These stones originate from volcanic action being formed by the crystallization of molten rock matter derived from deep in the earth's crust. It is the proportions of these crystals which give the stones their colour and characteristics. Granites are typical of this class of stone being hard, durable and capable of a fine, polished finish. Granites are mainly composed of quartz, felspar and mica.

Sedimentary stones

These stones are largely composed of material derived from the breakdown and erosion of existing rocks deposited in layers under the waters which, at that time, covered much of the earth's surface. Being deposited in this manner their section is stratified and shows to a lesser or greater degree the layers as deposited. Some of these layers are only visible if viewed under a microscope. Under the microscope it will be seen that all the particles lie in one direction, indicating the flow of the current in the water. Sandstones and limestones are typical examples of sedimentary stones.

Sandstones are stratified sedimentary rocks produced by eroded and disintegrated rocks, like granite, being carried away and deposited by water in layers. The brown and yellow tints in sandstones are due to the presence of oxides of iron.

Limestones may be organically formed by the deposit of tiny shells and calcareous (containing limestone) skeletons in the seas and rivers, or it may be formed chemically by deposits of lime in ringed layers. Limestones vary considerably from a heavy crystalline form to a friable material such as chalk.

Metamorphic stones

These are stones which have altered and may have been originally igneous or sedimentary rocks, but have since been changed by geological processes such as pressure, movement, heat and chemical reaction due to the infiltration of fluids. Typical examples of this type of stone are marbles and slates.

Marbles are metamorphic limestones, their original structure having been changed by pressure. Marbles, being capable of taking a high polish, are used mainly for decorative work.

Slate is a metamorphic clay having been subjected to great pressure and heat; being derived from a sedimentary layer it can be easily split into thin members.

Stones are obtained from quarries by blasting and wedging the blocks away from the solid mass. They are partly worked in the quarry and then sent to store yards where they can be sawn, cut, moulded, dressed and polished to the customer's requirements.

Today, natural stones are sometimes used for facing prestige buildings, constructing boundary or similar walls and in those areas where natural stones occur, to preserve the character of the district. The distribution of natural stones in the British Isles is shown in Fig. 3.1.1.

Substitutes for natural stones are available in the form of cast stones either as reconstructed or artificial stones. BS 1217 defines these as a building material manufactured from cement and natural aggregate, for use in a manner similar to and for the same purpose as natural stone.

Reconstructed stone

These types of cast stone are homogeneous throughout and therefore have the same texture and colour as the natural stones they are intended to substitute. They are free from flaws and stratification and can be worked in the same manner as natural stone or, alternatively, they can be cast into shaped moulds giving the required section.

Artificial stone

These stones consist partly of a facing material and partly of a structural concrete. The facing is a mixture of fine aggregate of natural stone and cement or sand and

(dotted)	Igneous
(blank)	Metamorphic and old sedimentary
(grid)	Old sandstones
(vertical lines)	New sandstones
(diagonal lines)	Limestones
(horizontal lines)	Chalk
(diagonal lines)	Clays

Figure 3.1.1 Distribution of natural stones

pigmented cement to resemble the natural stone colouring. This facing should be cast as an integral part of the stone and have a minimum thickness of 20 mm. They are cheaper than reconstructed stones but have the disadvantage that, if damaged, the concrete core may be exposed.

Stonework terminology

- **Arris**   Meeting edge of two worked surfaces.
- **Ashlar**   A square hewn stone; stonework consisting of blocks of stone finely squared and dressed to given dimensions and laid to courses of not less than 300 mm in height.
- **Bed joint**   Horizontal joint between two consecutive courses.
- **Bonders**   Through stones or stones penetrating two-thirds of the thickness of a wall.
- **Cramp**   Non-ferrous metal or slate tie across a joint.
- **Dowel**   Non-ferrous or slate peg morticed into adjacent joints.
- **Joggel**   Recessed key filled with a suitable material, used between adjacent vertical joints.
- **Lacing**   Course of different material to add strength.
- **Natural bed**   Plane of stratification in sedimentary stones.
- **Quarry sap**   Moisture contained in newly quarried stones.
- **Quoin**   Corner stone.
- **Stool**   Flat seating on a weathered sill for jamb or mullion.
- **String course**   Distinctive course or band used mainly for decoration.
- **Weathering**   Sloping surface to part of the structure to help shed the rain.

Ashlar walling

This form of stone walling is composed of carefully worked stones, regularly coursed, bonded and set with thin or rusticated joints and is used for the majority of high-class facing work in stone. The quoins are sometimes given a surface treatment to emphasize the opening or corner of the building. The majority of ashlar work is carried out in limestone varying in thickness from 100 to 300 mm and set in mason's putty, which is a mixture of stone dust, lime putty and Portland cement, a typical mix ratio being 7:5:2.

Rules for ashlar work

1. Back faces of ashlar stones should be painted with a bituminous or similar waterproofing paint.
2. External stonework must not be taken through the thickness of the wall since this could create a passage for moisture.
3. Ledges of cornices and external projections should be covered with lead, copper or asphalt to prevent damage by rain or birds.
4. Moulded cornices should be raked back at 45° to counteract the cantilever action.
5. Face of stones should be given a protective coat of slurry during construction, the slurry being washed off immediately prior to completion.

Typical details of ashlar work are shown in Figs 3.1.2–3.1.5.

Alternative cornice detail showing joints

Figure 3.1.2   Typical details of ashlar stonework

Figure 3.1.3   Typical surface and joint treatments

Figure 3.1.4 Typical treatments to arch openings

Figure 3.1.5   Typical treatments to square openings

Rubble walling

These are walls consisting of stones which are left in a rough or uneven state thus presenting a natural appearance to the face of the wall. These stones are usually laid with a wide joint and are frequently used in various forms in many rural areas. They can be laid dry or bedded in earth in boundary walls or bedded in lime mortar when used for the walls of farm outbuildings; if used in conjunction with ashlar stonework, a cement or gauged mortar is used. It is usual to build the quoins to corners, window and door openings in dressed or ashlar stones. As with ashlar work it is advisable to treat the face of any backing material with a suitable water-proofing coat to prevent the passage of moisture or the appearance of cement stains on the stone face.

Solid stone walls will behave in the same manner as solid brick walls with regard to the penetration of moisture and rain, it will therefore be necessary to take the same precautions in the form of damp-proof courses to comply with Part C of the Building Regulations.

Typical examples of stone and rubble walling are shown in Figs 3.1.4, 3.1.6 and 3.1.7.

BRICKWORK

The history of the art of brickmaking and the craft of bricklaying can be traced back to before 6,000 BC. It started in western Asia, spread eastwards and was introduced into this country by the Romans. The use of brickwork flourished during the third and fourth centuries, after which the craft suffered a rapid decline until its reintroduction from Flanders towards the end of the fourteenth century. Since then it has been firmly established and remains as one of the major building materials.

BS 3921 specifically relates to bricks manufactured from clay for use in walling. Other standards exist for units produced from calcium silicate (BS 187) and concrete (BS 6073). A brick is defined in BS 3921 as a walling unit with co-ordinating or format size of 225 mm length, 112.5 mm width and 75 mm height. Bricks are known by their format size – that is, the actual or work size plus a 10 mm mortar joint allowance to three faces. Therefore, the standard brick of $225 \times 112.5 \times 75$ mm has actual dimensions of $215 \times 102.5 \times 65$ mm. The terms used for bricks and brickwork are shown in Figs 3.1.8, 3.1.9 and 3.1.10.

Brickwork is used primarily in the construction of walls by the bedding and jointing of bricks into established bonding arrangements. The term also covers the building in of hollow and other lightweight blocks. The majority of the bricks used today are made from clay or shale conforming to the requirements of BS 3921.

MANUFACTURE OF CLAY BRICKS

The basic raw material is clay, shale or brickearth of which this country has a good supply. The raw material is dug and then prepared either by weathering or grinding

Rubble coping bedded
in cement mortar

Random rubble set
dry or bedded in
cement mortar

Gaps filled with
spalls or small
stones

Bonder or
through stone
1 per m²

Foundation of mass concrete
or large stones

Uncoursed random rubble wall

Weathered stone coping
bedded in cement mortar

Random rubble
bedded in cement
mortar and laid
in courses 225
to 450 high

Wide joints

Foundation of mass
concrete or large stones

Coursed random rubble wall

Figure 3.1.6   Typical rubble walls

Segmental stone coping
bedded in cement mortar

Roughly square
rubble bedded in
cement mortar
and laid in courses
300 to 450 high

Thin
joints

Foundation of mass
concrete or large stones

Coursed square rubble wall

Saddle back stone coping
bedded in cement mortar

Flints from the
lower regions of
the chalk strata
bedded in
cement mortar

Brick lacing
courses for
added strength

Foundation of mass
concrete or large stones

Flint stone wall

Figure 3.1.7   Typical stone walls

Figure 3.1.8   Brickwork terminology

Figure 3.1.9   Standard brick

Figure 3.1.10   Keyed brick

before being mixed with water to the right plastic condition. It is then formed into the required brick shape before being dried and fired in a kiln.

Different clays have different characteristics, such as moisture content and chemical composition, therefore distinct variations of the broad manufacturing processes have been developed and these are easily recognized by the finished product.

Pressed bricks

This is the type of brick most commonly used, accounting for nearly two-thirds of the 8,400 million produced in this country each year. There are two processes of pressed brick manufacture: the semi-dry method and the stiff plastic method.

By far the greatest number of bricks are made by the **semi-dry pressed process** and are called 'flettons'; these form over 40% of the total brick production in Britain. The name originates from the village of Fletton outside Peterborough where the bricks were first made. This process is used for the manufacture of bricks from the Oxford clays which have a low natural plasticity. The clay is ground, screened and pressed directly into the moulds.

The **stiff plastic process** is used mainly in Scotland, the north of England and south Wales. The clays in these areas require more grinding than the Oxford clays and the clay dust needs tempering (mixing with water) before being pressed into the mould.

All pressed bricks contain frogs which are sometimes pressed on both bed faces. In general, pressed bricks are more accurate in shape than other clay bricks with sharp arrises and plain faces.

Wire cut bricks

Approximately 28% of bricks produced in Britain are made by this process. The clay, which is usually fairly soft and of a fine texture, is extruded as a continuous ribbon and is cut into brick units by tightly stretched wires spaced at the height or depth for the required brick. Allowance is made during the extrusion and cutting for the shrinkage that will occur during firing. Wire cut bricks do not have frogs and on many the wire cutting marks can be clearly seen.

Soft mud process bricks

This process is confined mainly to the south-eastern and counties of England where suitable soft clays are found. The manufacture can be carried out by machine or by hand either with the natural clay or with a mixture of clay and lime or chalk. In both methods the brick is usually frogged and is less accurate in shape than other forms of bricks. Sand is usually used in the moulds to enable the bricks to be easily removed, and this causes an uneven patterning or creasing on the face.

BRICK CLASSIFICATION

Bricks are a diverse product, therefore they are difficult to standardize for classification. They may be known by the terms used in BS 3921 – that is, compressive strength (N/mm^2) and percentage by mass water absorption. Alternatively, the manufacturer's name or catalogue descriptions may be used, possibly in combination with the BS characteristics. The minimum compressive strength acceptable to the Building Regulations is 5 N/mm^2 over gross sectional area (2.8 N/mm^2 for blocks).

Some common terminology:

Varieties

Common   Suitable for general building work but having no special claim to give an attractive appearance.

Facing   Specially made or selected to have an attractive appearance when used without rendering or plaster.

Engineering   Having a dense and strong semi-vitreous body conforming to defined limits for absorption and strength.

Qualities

Internal   Suitable for internal use only, may need protection on site during bad weather or during the winter.

Ordinary   Less durable than special quality but normally durable in the external face of a building. Some types are unsuitable for exposed situations.

Special   For use in conditions of extreme exposure where the structure may become saturated and frozen such as retaining walls and pavings.

Types

Solid   Those in which small holes passing through or nearly through the brick do not exceed 25% of its volume or in which frogs do not exceed 20% of its volume. A small hole is defined as a hole less than 20 mm wide or less than 500 mm^2 in area.

Perforated   Those in which holes passing through the brick exceed 25% of its volume and the holes are small as defined above.

Hollow   Those in which the holes passing through the brick exceed 25% of its volume and the holes are larger than those defined as small holes.

Cellular   Those in which the holes are closed at one end and exceed 20% of the volume of the brick.

Other classifications

Bricks may also be classified by one or more of the following:

- Place of origin, for example, London.
- Raw material, for example, clay.
- Manufacture, for example, wire cut.
- Use, for example, foundation.
- Colour, for example, blue.
- Surface texture, for example, sand-faced.

CALCIUM SILICATE BRICKS

These bricks are also called sandlime and sometimes flintlime bricks and are covered by BS 187, which gives several classes of brick; the higher the numbered class, the stronger is the brick. The format size of a calcium silicate brick is the same as that given for a standard clay brick.

These bricks are made from carefully selected clean sand and/or crushed flint mixed with controlled quantities of lime and water. At this stage colouring pigments can be added if required; the relatively dry mix is then fed into presses to be formed into the required shape. The moulded bricks are then hardened in sealed and steam-pressurized autoclaves. This process, which takes from seven to ten hours, causes a reaction between the sand and the lime resulting in a strong homogeneous brick which is ready for immediate delivery and laying. The bricks are very accurate in size and shape but do not have the individual characteristics of clay bricks.

CONCRETE BRICKS

These are made from a mixture of inert aggregate and cement in a similar fashion to calcium silicate bricks and are cured either by natural weathering or in an autoclave. Details of the types and properties available as standard concrete bricks are given in BS 6073.

MORTARS FOR BRICKWORK

The mortar used in brickwork transfers the tensile, compressive and shear stresses uniformly between adjacent bricks. To do this it must satisfy certain requirements:

1. Have adequate strength, but not greater than that required for the design strength.
2. Have good workability.
3. Needs to retain plasticity long enough for the bricks to be laid.
4. Be durable over a long period.
5. Bond well to the bricks.
6. Be able to be produced at an economic cost.

If the mortar is weaker than the bricks shrinkage cracks will tend to follow the joints of the brickwork and these are reasonably easy to make good. If the mortar is stronger than the bricks shrinkage cracks will tend to be vertical through the joints and the bricks thus weakening the fabric of the structure.

Mortar mixes

Mortar is a mixture of sand and lime or a mixture of sand and cement with or without lime. Proportioning of the materials can be carried by volume but this method is inaccurate and it is much better to proportion by weight. The effect of the lime is to make the mix more workable, but as the lime content increases the mortar's resistance to damage by frost action decreases.

Plasticizers by having the effect of entraining small bubbles of air in the mix and breaking down surface tension will also increase the workability of a mortar.

Mortars should never be re-tempered and should be used within two hours of mixing or be discarded.

Typical mixes (by volume)

Cement mortar (cement:sand)
- 1:3 – suitable for brickwork in exposed conditions such as parapets and for brickwork in foundations.

Lime mortar (lime:sand)
- 1:3 – for internal use only.

Gauged mortars (cement:lime:sand)
- 1:1:6 – suitable for most conditions of severe exposure.
- 1:2:9 – suitable for most conditions except those of severe exposure.
- 1:3:12 – internal use only.

DAMPNESS PENETRATION

It is possible for dampness to penetrate into a building through the brick walls by one or more of three ways:

1. By the rain penetrating the head of the wall and soaking down into the building below the roof level.
2. By the rain beating against the external wall and soaking through the fabric into the building.
3. By ground moisture entering the wall at or near to the base and creeping up the wall by capillary action and entering the building above the ground floor level.

Numbers 1 and 3 can be overcome by the insertion of a suitable damp-proof course in the thickness of the wall. Number 2 can be overcome by one of two methods:

■ Applying to the exposed face of the wall a barrier such as cement rendering or some suitable cladding like vertical tile hanging.
■ Constructing a cavity wall, whereby only the external skin becomes damp, the cavity providing a suitable barrier to the passage of moisture through the wall.

Damp-proof courses

The purpose of a damp-proof course in a building is to provide a barrier to the passage of moisture from an external source into the fabric of the building or from one part of the structure to another. Damp-proof courses may be either horizontal or vertical and can generally be divided into three types:

1. Those below ground level to prevent the entry of moisture from the soil.
2. Those placed just above ground level to prevent moisture creeping up the wall by capillary action; this is sometimes called **rising damp**.
3. Those placed at openings, parapets and similar locations to exclude the entry of the rainwater which falls directly onto the fabric of the structure.

Materials for damp-proof courses

BS 743: *Specification for materials for DPCs* gives seven suitable materials for the construction of damp-proof courses, all of which should have the following properties:

■ Be completely impervious.
■ Be durable, having a longer life than the other components in the building and therefore should not need replacing during its lifetime.
■ Be in comparatively thin sheets so as to prevent disfigurement of the building.
■ Be strong enough to support the loads placed upon it without exuding from the wall.
■ Be flexible enough to give with any settlement of the building without fracturing.

The following also apply:

■ BS 6398: *Specification for bitumen damp proof courses for masonry.*
■ BS 6515: *Specification for polythene damp proof courses for masonry.*
■ BS 8215: *Code of practice for design and installation of damp proof courses in masonry construction.*

Lead   This should be at least code No. 4–BS 1178 lead; it is a flexible material supplied in thin sheets and therefore large irregular shapes with few joints can be formed, but it has the disadvantages of being expensive and liable to exude under heavy loadings.

Copper   This should have a minimum thickness of 0.25 mm; like lead it is supplied in thin sheets and is expensive.

Figure 3.1.11 Special bricks

Bitumen This is supplied in the form of a felt usually to brick widths and is therefore laid quickly with the minimum number of joints. Various bases are available, such as hessian, fibre, asbestos and lead, all of which are inexpensive but have the main disadvantage of being easily torn.

Mastic asphalt Applied in two layers giving a total thickness of 25 mm, it is applied *in-situ* and is therefore jointless but is expensive in small quantities.

Polythene Black low-density polythene sheet of single thickness not less than 0.5 mm thick should be used; it is easily laid but can be torn and punctured easily.

Slates These should not be less than 230 mm long nor less than 4 mm thick and laid in two courses set breaking the joint in cement mortar 1:3. Slates have limited flexibility but are impervious and very durable. Cost depends upon the area in which the building is being erected.

Bricks These should comply with the requirements of BS 3921 'engineering' classification. They are laid in two courses in cement mortar and can be out of context with the general façade of the building.

BRICKWORK BONDING

When building with bricks it is necessary to lay the bricks to some recognized pattern or bond in order to ensure stability of the structure and to produce a pleasing appearance. All the various bonds are designed so that no vertical joint in any one course is directly above or below a vertical joint in the adjoining course. To simplify this requirement special bricks are produced to BS 4729: *Specification for dimensions of bricks of special shapes and sizes.* Alternatively, the bricklayer can cut from whole bricks on site. Application of some of these specials is shown in Fig. 3.1.11 with some additions in Fig. 3.1.12. The various bonds are also planned to give the greatest practical amount of lap to all the bricks and this should not be less than a quarter of a brick length. Properly bonded brickwork distributes the load over as large an area of brickwork as possible, then angle of spread of the load through the bonded brickwork is 60°.

Common bonds

Stretcher bond Consists of all stretchers in every course and is used for half-brick walls and the half-brick skins of hollow or cavity walls (see Fig. 3.1.13).

English bond A very strong bond consisting of alternate courses of headers and stretchers (see Fig. 3.1.14).

Flemish bond Each course consists of alternate headers and stretchers, its appearance is considered to be better than English bond but it is not quite so

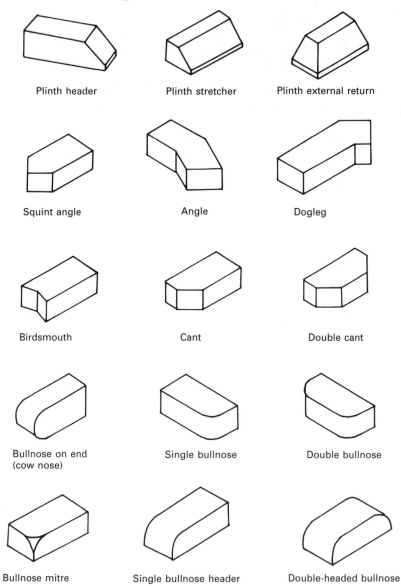

Plinth header   Plinth stretcher   Plinth external return

Squint angle   Angle   Dogleg

Birdsmouth   Cant   Double cant

Bullnose on end   Single bullnose   Double bullnose
(cow nose)

Bullnose mitre   Single bullnose header   Double-headed bullnose

Figure 3.1.12   Purpose-made bricks

strong. This bond requires fewer facing bricks than English bond, needing only 79 bricks per square metre as opposed to 89 facing bricks per square metre for English bond. This bond is sometimes referred to as double Flemish bond (see Fig. 3.1.15).

Single Flemish bond   A combination of English and Flemish bonds having Flemish bond on the front face with a backing of English bond. It is considered to

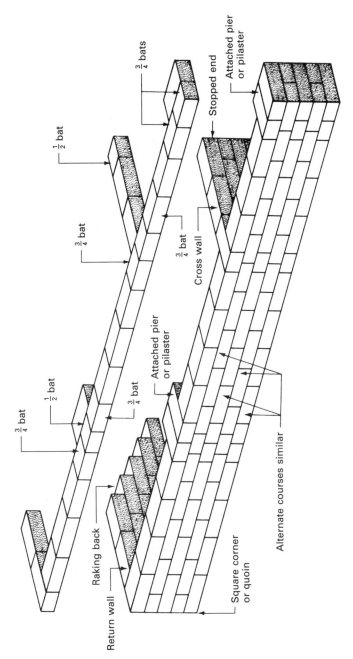

¾ bat

½ bat

¾ bat

¾ bat

Cross wall

Attached pier
or pilaster

Stopped end

Attached pier
or pilaster

¾ bats

½ bat

Return wall

Raking back

¾ bat

½ bat

¾ bat

Attached pier
or pilaster

Alternate courses similar

Square corner
or quoin

Figure 3.1.13 Typical stretcher bond details

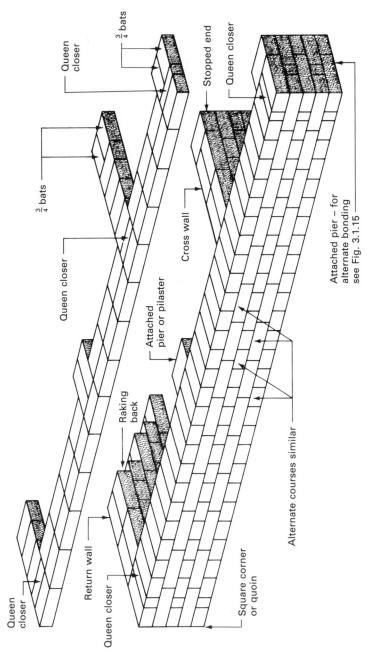

Queen closer

Return wall

Queen closer

Raking back

Attached pier or pilaster

Queen closer

Cross wall

$\frac{3}{4}$ bats

Queen closer

$\frac{3}{4}$ bats

Stopped end

Queen closer

Attached pier – for alternate bonding see Fig. 3.1.15

Alternate courses similar

Square corner or quoin

Figure 3.1.14 Typical English bond details

Queen closer

Return wall

Queen closer

Raking back

$\frac{1}{2}$ bat

Quoin

Attached
pier or
pilaster

$\frac{3}{4}$ bats

$\frac{3}{4}$ bats

Stopped end

Cross wall

Attached
pier or
pilaster

Queen
closer

Queen
closer

Alternate courses similar

Attached pier – for
alternative bonding
see Fig. 3.1.14

Note: The above example shows reverse bond, i.e. the
end bricks in each course do not correspond –
this technique can be used with all other bonds

Figure 3.1.15 Typical Flemish bond details

be slightly stronger than Flemish bond. The thinnest wall that can be built using this bond is a one-and-a-half brick wall.

English garden wall bond   Consists of three courses of stretchers to one course of headers.

Flemish garden wall bond   Consists of one header to every three stretchers in every course; this bond is fairly economical in facing bricks and has a pleasing appearance.

Special bonds

Rat-trap bond   This is a brick on edge bond and gives a saving on materials and loadings, suitable as a backing wall to a cladding such as tile hanging (see Fig. 3.1.16).

Quetta bond   Used on one-and-a-half brick walls for added strength, suitable for retaining walls (see Fig. 3.1.16).

Metric modular brickwork

The standard imperial format brick does not fit reasonably well into the system of dimensional co-ordination, with its preferred dimension of 300 mm. Metric modular bricks have been designed and produced with four different formats (see Fig. 3.1.17), but BS 6649: *Specification for clay and calcium silicate modular bricks*, provides only for a single format of $200 \times 100 \times 75$ mm.

The bond arrangements are similar to the well-known bonds but are based on third bonding, that is the overlap is one-third of a brick and not one-quarter as with the standard format brick. Examples of metric modular brick bonding are shown in Fig. 3.1.17.

Footings

These are wide courses of bricks placed at the base of a wall to spread the load over a greater area of the foundations. This method is seldom used today, instead the concrete foundation would be reinforced to act as a beam or reinforced strip. The courses in footings are always laid as headers as far as possible; stretchers, if needed, are laid in the centre of the wall (see Fig. 3.1.18).

Boundary walls

These are subjected to severe weather conditions and therefore should be correctly designed and constructed. If these walls are also acting as a retaining wall the

Figure 3.1.16 Special bonds

conditions will be even more extreme, but the main design principle of the exclusion of water remains the same. The presence of water in brickwork can lead to frost damage, mortar failure and efflorescence. The incorporation of adequate damp-proof courses and overhanging throated copings is of the utmost importance in this form of structure (see Fig. 3.1.19).

Metric modular bricks

Figure 3.1.17   Metric modular brickwork

Figure 3.1.18 Typical footings

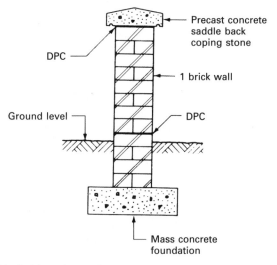

Figure 3.1.19 Typical boundary wall

Efflorescence

This is a white stain appearing on the face of brickwork caused by deposits of soluble salts formed on or near the surface of the brickwork as a result of evaporation of the water in which they have been dissolved. It is usually harmless and disappears within a short period of time; dry brushing or with clean water may be used to remove the salt deposit but the use of acids should be left to the expert.

BLOCKWORK

A block can be defined as a walling unit exceeding the BS dimensions specified for bricks and its height shall not exceed either its length or six times its thickness. This avoids confusion with slabs or panels. Blocks are produced in clay and various cement/aggregate mixes, including wood chippings.

CLAY BLOCKS

These have format size of $300 \times 225 \times 62.5$, 75, 100 or 150 mm width. They are manufactured hollow by an extrusion process and are fired as for clay bricks. The standard six (150 mm) cavity block has been used for the inner skin of cavity walls, but has now been superseded by developments in lightweight concrete blocks. The three (75 mm) cavity block is primarily intended for partitions. 150 mm blocks are also used in the warmer climates of the world as a single leaf external walling with rendered and emulsioned finish. Chapter 4 shows another application to structural concrete floors. Typical details are shown in Fig. 3.1.20.

PRECAST CONCRETE BLOCKS

The specification and manufacture of precast concrete blocks or masonry units is covered in BS 6073. Classification is by compressive strength categories: 2.8, 3.5, 5, 7, 10, 15, 20 and 35 N/mm^2.

The density of a precast concrete block gives an indication of its compressive strength – the greater the density the stronger is the block. Density will also give an indication as to the thermal conductivity and acoustic properties of a block. The lower the density the lower is the thermal conductivity factor, whereas the higher the density the greater is the reduction of airborne sound through the block.

The actual properties of different types of precast concrete block can be obtained from manufacturers' literature together with their appearance classification such as plain, facing or special facing.

Aerated concrete for blocks is produced by introducing air or gas into the mix so that, when set, a uniform cellular block is formed. The usual method employed is to introduce a controlled amount of fine aluminium powder into the mix which reacts with the free lime in the cement to give off hydrogen which is quickly replaced by air and so provides the aeration.

Precast concrete blocks are manufactured to a wide range of standard sizes, the most common face format sizes being 400×200 mm and 450×225 mm, with thicknesses of 75, 90, 100, 140, 190 and 215 mm. Typical details are shown in Fig. 3.1.21.

Concrete blocks are laid in what is essentially stretcher bond and joined to other walls by block bonding or leaving metal ties or strips projecting from suitable bed courses. As with brickwork the mortar used in blockwork should be weaker than the material of the walling unit, generally a 1:2:9 gauged mortar mix will be suitable for work above ground level.

Figure 3.1.20 Hollow clay blocks and blockwork

Concrete blocks shrink on drying out, therefore they should not be laid until the initial drying shrinkage has taken place (usually this is about fourteen days under normal drying conditions) and should be protected on site to prevent them becoming wet, expanding and causing subsequent shrinkage possibly resulting in

Figure 3.1.21   Precast concrete blocks and blockwork

cracking of the blocks and any applied finishes such as plaster. Where length of wall exceed 6 metres or thereabouts, it is necessary to incorporate vertical movement joints. These comprise a stainless steel or galvanized steel former incorporating horizontal bed joint ties at 450 mm spacing. Figure 3.1.22 shows the installation with bonded profiled or perforated ties to one side and plastic sleeved ties to the other to maintain continuity and facilitate movement.

Ribbed bonding tie

Former

Plastic sleeved continuity tie

Galvanized or stainless steel
former at 6 m spacing

Tie

Ties at 450 mm
spacing, initial tie
at 225 mm

Lightweight concrete
blockwork partition
wall

Mastic sealant
pointing and fire
stopping (if required)

Figure 3.1.22   Blockwork movement joint (BS 5628: Part 3: *Use of masonry*)

The main advantages of blockwork over brickwork are:

■ Labour saving – easy to cut, larger units.
■ Easier fixings – most take direct fixing of screws and nails.
■ Higher thermal insulation properties.
■ Lower density.
■ Provide a suitable key for plaster and cement rendering.

The main disadvantages are:

■ Lower strength.
■ Less resistance to rain penetration.
■ Loadbearing properties less (one- or two-storey application).
■ Lower sound insulation properties.

CAVITY WALLS

A wall constructed in two leaves or skins with a space or cavity between them is called a cavity wall and it is the most common form of external wall used in domestic building today. The main purpose of constructing a cavity wall is to prevent the penetration of rain to the internal surface of the wall. It is essential that the cavity is not bridged in any way as this would provide a passage for the moisture.

Air bricks have been used to ventilate the cavity and these would be built in at the head and base of the cavity wall in order that a flow of air may pass through the cavity thus drying out any moisture that has penetrated the outer leaf. Unless the wall is exposed to very wet conditions the practice of inserting air bricks to ventilate the cavity is not recommended since it lowers the thermal and sound insulation values of the wall.

The main consideration in the construction of a cavity wall above ground level damp-proof course is the choice of a brick or block which will give the required durability, strength and appearance and also conform to Building Regulation requirements. The main function of the wall below ground level damp-proof course is to transmit the load safely to the foundations, in this context the two half-brick or block leaves forming the wall act as retaining walls. There is a tendency for the two leaves to move towards each other due to the pressure of the soil and the space provided by the cavity. To overcome this problem the cavity below ground level is filled with a weak mix of concrete, thus creating a solid wall in the ground (see Fig. 3.2.1). Alternatively, the substructural wall may be built solid. It is also advisable to leave out every fourth vertical joint in the external leaf at the base of the cavity and above the cavity fill, to allow any moisture trapped in the cavity a means of escape.

Parapets, whether solid or cavity construction, are exposed to the elements on three sides and need careful design and construction. They must be provided with adequate barriers to moisture in the form of damp-proof courses since dampness could penetrate the structure by soaking down the wall and bypassing the roof and entering the building below the uppermost ceiling level. A solid parapet wall should not be less than 150 mm thick and not less than the thickness of the wall

Outline
of roof

6 mm cement particle
board cavity closer

Block inner leaf (90 mm min.)

Facing
bricks
(102.5 mm)

Galvanized steel lintel
with insulation

Direct glazed
window frame

Internal reveal
usually plastered
or lined

Brick
external reveal

Window board

Wall tie

50 or 65 mm fully insulated cavity

DPC

Solid ground floor

150 mm
min.

DPM over sand
blinding and hardcore

Open joints
at 900 c/c

Loadbearing concrete
blocks or bricks
below ground

Weak concrete
cavity filling

Mass concrete
strip foundation

Figure 3.2.1 Typical cavity wall details

on which it is carried and its height should not exceed four times its thickness.
The recommended maximum heights of cavity wall parapets is shown in Fig. 3.2.2.

Building Regulations

Regulation A1 requires that a building shall be so constructed that the combined
dead, imposed and wind loads are sustained and transmitted to the ground safely
and without causing any movement which will impair the stability of any part of

Sum of leaf thicknesses (mm)	Height (mm)
≤ 200	600 max.
200–250	860 max.

Figure 3.2.2 Parapets

another building. Guidance to meet these requirements for cavity walls is given in Approved Document A.

Section 2C of this document considers full storey height cavity walls for residential buildings of up to three storeys and requires that:

1. The compressive strengths of bricks and blocks should be not less than 5 N/mm² and 2.8 N/mm² respectively for buildings up to two storeys. Greater strength classifications are required for three storey buildings (see Approved Document A, Section 2C, Diagram 7).
2. Cavities (gap between masonry faces) should be at least 50 mm, but may be up to 300 mm. Both leaves in cavity walls, should have ties embedded at least 50 mm into adjacent masonry.
3. Wall ties should comply with BS 1243: *Specification for metal ties for cavity wall construction*. Plastic and galvanised steel ties have been used, but now the preferred material is austenitic stainless steel. Maximum spacings are shown in Figure 3.2.3. Cavities < 75 mm may be tied with butterfly pattern ties but in excess of this they should be provided with twisted pattern ties.
4. Cavity walls normally have leaves of at least 90 mm thick. An exception is with leaves of 65–90 mm thickness and cavity up to 75 mm with ties at 450 mm max. horizontal spacing.
5. The combined thickness of the two leaves of a cavity wall should be not less than 190 mm for a maximum wall height of 3.5 m and length not exceeding 12 m

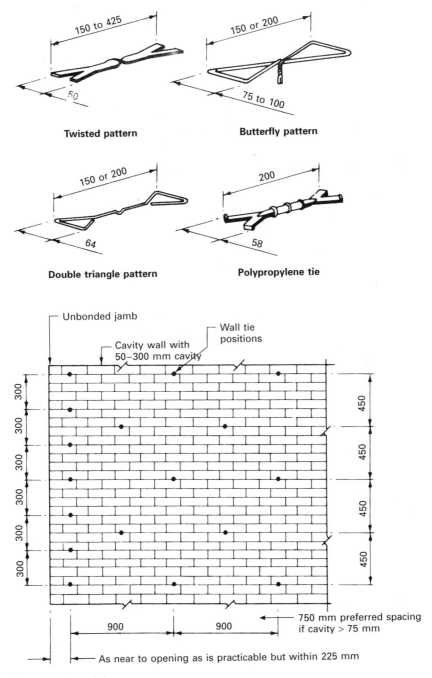

150 to 425

50

Twisted pattern

150 or 200

75 to 100

Butterfly pattern

150 or 200

64

Double triangle pattern

200

58

Polypropylene tie

Unbonded jamb

Cavity wall with
50–300 mm cavity

Wall tie
positions

300

300

300

300

300

300

450

450

450

450

900

900

750 mm preferred spacing
if cavity > 75 mm

As near to opening as is practicable but within 225 mm

Figure 3.2.3 Wall ties

and also for heights between 3.5 m and 9 m in wall lengths not exceeding 9 m. Wall lengths and heights up to 12 m require a minimum thickness of 290 mm. Variations:

Wall height	Wall length	Wall thickness
≤ 12 m	≤ 9 m	290 mm for one storey
		190 mm for remainder
≤ 12 m	≤ 12 m	290 mm for 2 storeys
		190 mm for remainder

6. Mortar should be as given for mortar designation (iii) in BS 5628–1: *Structural use of unreinforced masonry*, or a gauged mortar mix of 1:1:6 by volume, or its equivalent.
7. Cavity walls of any length need to be provided with roof lateral support. Walls over 3 m length will also require floor lateral support at every floor forming a junction with the supported wall. If roof lateral support is not provided by type of covering (tiles or slates), a pitch of 15° or more plus a minimum wall plate bearing of 75 mm, durable metal straps with a minimum cross-section of 30 × 5 mm will be needed at not more than 2 m centres (see Fig. 6.1.9). If the floor does not have at least a 90 mm bearing on the supported wall, lateral support should be provided by similar straps at not more than 2 m spacing or the joists should be fixed using restraint-type joist hangers (see Fig. 4.5.1).

Prevention of damp in cavity walls

Approved Document C recommends a cavity to be carried down at least 150 mm below the lowest damp-proof course and that any bridging of the cavity, other than a wall tie or closing course protected by the roof, is to have a suitable damp-proof course to prevent the passage of moisture across the cavity. Where the cavity is closed at the jambs of openings a vertical damp-proof course should be inserted unless some other suitable method is used to prevent the passage of moisture from the outer leaf to the inner leaf of the wall.

Approved Document C recommends a damp-proof course to be inserted in all external walls at least 150 mm above the highest adjoining ground or paving to prevent the passage of moisture rising up the wall and into the building, unless the design is such that the wall is protected or sheltered.

Advantages of a cavity in wall construction

These can be listed as follows:

- Able to withstand a driving rain in all situations from penetrating to the inner wall surface.
- Gives good thermal insulation, keeping the building warm in winter and cool in the summer.
- No need for external rendering.

Figure 3.2.4 Brick joints

- Enables the use of cheaper and alternative materials for the inner construction.
- A nominal 255 mm cavity wall has a higher sound insulation value than a standard one brick thick wall.
- Can accommodate supplementary insulation material.

Disadvantages of cavity wall construction

These can be listed as follows:

- Requires a high standard of design and workmanship to produce a soundly constructed wall; this will require good supervision during construction.
- The need to include a vertical damp-proof course to all openings.
- Slightly dearer in cost than a standard one brick thick wall.

JOINTING AND POINTING

These terms are used for the finish given to both the vertical and horizontal joints in brickwork irrespective of whether the wall is of brick, block, solid or cavity construction.

Jointing is the finish given to the joints when carried out as the work proceeds.

Pointing is the finish given to the joints by raking out to depth of approximately 20 mm and filling in on the face with a hard setting cement mortar which could have a colour additive. This process can be applied to both new and old buildings. Typical examples of jointing and pointing are shown in Fig. 3.2.4.

OPENINGS IN WALLS

An opening in an external wall consists of a head, jambs or reveals and a sill or threshold.

HEAD

The function of a head is to carry the load of brickwork over the opening and transmit this load to the jambs at the sides. To fulfil this task it must have the capacity to support the load without unacceptable deflection. A variety of materials and methods is available in the form of a lintel or beam such as:

Timber   Suitable for light loads and small spans, the timber should be treated with a preservative to prevent attack by beetles or fungi.

Steel

- *For small openings* – a mild steel flat or angle section can be used to carry the outside leaf of a cavity wall, the inner leaf being supported by a concrete or steel lintel.
- *For medium spans* – a channel or joist section is usually suitable.
- *For large spans* – a universal beam section to design calculations will be needed.

Steel lintels and beams which are exposed to the elements should be either galvanized or painted with several coats of bituminous paint to give them protection against corrosion.

Concrete   These can be designed as *in-situ* or precast reinforced beams or lintels and can be used for all spans. Prestressed concrete lintels are available for the small and medium spans (see Chapter 10, *Advanced Construction Technology*).

Stone   These can be natural, artificial or reconstructed stone but are generally used as a facing to a steel or concrete lintel (see Fig. 3.1.5).

Brick   Unless reinforced with mild steel bars or mesh, brick lintels are only suitable for small spans up to 1 m, but like stone, bricks are also employed as a facing to a steel or concrete lintel.

Lintels require a bearing at each end of the opening; the amount will vary with the span but generally it will be 100 mm for the small spans and up to 225 mm for the medium and large spans. In cavity walling a damp-proof course will be required where the cavity is bridged by the lintel and this should extend at least 150 mm beyond each end of the lintel. Open joints are sometimes used to act as weep holes; these are placed at 900 mm centres in the outer leaf immediately above the damp-proof course. Typical examples of head treatments to openings are shown in Figs 3.3.1 and 3.3.2.

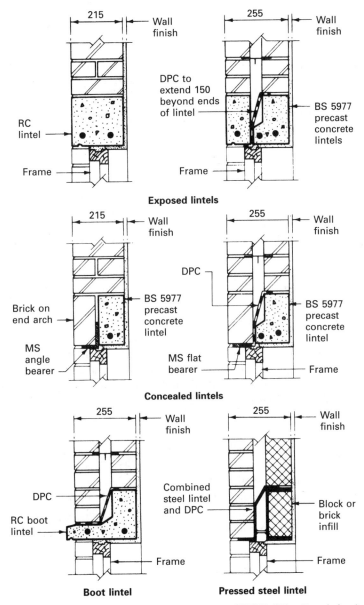

Figure 3.3.1 Typical head treatments to openings (pre-1985 Building Regulations)

Figure 3.3.2   Typical head treatments to openings

ARCHES

These are arrangements of wedge-shaped bricks designed to support each other and carry the load over the opening round a curved profile to abutments on either side. Full details of arch construction are given in the next chapter.

JAMBS

In solid walls these are bonded to give the required profile and strength; examples of bonded jambs are shown in Fig. 3.1.11. In cavity walls the cavity can be closed at the opening by using a suitable frame or by turning one of the leaves towards the other forming a butt joint in which is incorporated a vertical damp-proof course as required by the Building Regulations. Typical examples of jamb treatments to openings are shown in Figs 3.3.3 and 3.3.4.

SILL

The function of a sill is to shed the rain water, which has run down the face of the window or door and collected at the base, away from the opening and the face of the wall. Many methods and materials are available; appearance and durability are the main requirements since a sill is not a member which is

Wall finish

Closer

Bonded jamb
no vertical DPC
required

Frame

Wall finish

Wall ties at 300 c/c
for height of opening

Frame

Vertical DPC fixed to
back of frame

Wall finish

Wall ties of 300 c/c
for height of opening

Vertical DPC tacked
to back of frame and
bedded in cm. mt. (1:3)

Frame

Wall finish

Wall ties at 300 c/c
for height of opening

2 courses of slates
bedded in cm. mt. (1:3)
laid breaking joint

Frame

Figure 3.3.3 Typical jamb treatments to openings (pre-1985 Building Regulations)

Figure 3.3.4   Typical jamb treatments to openings

needed to carry heavy loads. Sills, unlike lintels, do not require a bearing at each end. Typical examples of sill treatments to openings are shown in Figs 3.3.5 and 3.3.6.

CONSERVATION OF FUEL ENERGY

The details shown in Figs 3.3.1, 3.3.3 and 3.3.5 are no longer acceptable by UK building standards for new housing, due to inadequate thermal insulation and the opportunity for thermal bridging (see Chapter 8.3). However, much of our existing housing stock is constructed in this manner, therefore these illustrations are retained as historic reference.

Figure 3.3.5 Typical sill treatments to openings (pre-1985 Building Regulations)

Hardwood sill

Window board

Insulation filler
under sill

Insulated
cavity
closer

Batt
retainer

Partial
fill cavity
insulation
batts

Wall tie
with disc
retainer

Cavity to suit
50 or 65 mm
standard insulation
batts

Cavity > 75 mm

Figure 3.3.6   Typical sill treatments to openings

ARCHES

These are arrangements of wedge-shaped bricks called *voussoirs* which are designed to support each other and carry the load over the opening, round a curved profile, to abutments on either side. An exception to this form is the flat or 'soldier' arch constructed of bricks laid on end or on edge.

When constructing an arch it must be given temporary support until the brick joints have set and the arch has gained sufficient strength to support itself and carry the load over the opening. These temporary supports are called centres and are usually made of timber; their design is governed by the span, load and thickness of the arch to be constructed.

SOLDIER ARCHES

This type of arch consists of a row of bricks showing on the face either the end or the edge of the bricks. Soldier arches have no real strength and if the span is over 1,000 mm they will require some form of permanent support such as a metal flat or angle (see Fig. 3.3.1). If permanent support is not given the load will be transferred to the head of the frame in the opening instead of the jambs on either side. Small spans can have an arch of bonded brickwork by inserting into the horizontal joints immediately above the opening some form of reinforcement such as expanded metal or bricktor, which is a woven strip of high tensile steel wires designed for the reinforcement of brick and stone walls. It is also possible to construct a soldier arch by inserting metal cramps in the vertical joints and casting these into an *in-situ* backing lintel of reinforced concrete.

ROUGH ARCHES

These arches are constructed of ordinary uncut bricks and being rectangular in shape they give rise to wedge-shaped joints. To prevent the thick end of the joint from becoming too excessive rough arches are usually constructed in header courses. The rough arch is used mainly as a backing or relieving arch to a gauged

Allow 40 mm for every 300 mm of span

Gauged brickwork

300

Allow 3 mm for
every 300 mm
of span

Camber to correct
illusion of sagging

Camber arch

Voussoirs

Key brick

Skewback

Abutment

Rise $\frac{1}{8}$ of
span

Span

Gauged segmental arch

Key brick

Voussoirs

Extrodos

Intrados or
soffit

Springing

Springer

Abutment

Gauged semi-circular arch

Figure 3.4.1   Typical examples of brick arches

brick or stone arch but they are sometimes used in facework for the cheaper form of building or where appearance is of little importance.

GAUGED ARCHES

These are the true arches and are constructed of bricks cut to the required wedge shape called voussoirs. The purpose of voussoirs is to produce a uniform thin joint which converges on to the centre point or points of the arch. There are two methods of cutting the bricks to the required wedge shape, namely, axed and rubbed. If the brick is of a hard nature it is first marked with a tin saw, to produce a sharp arris, and then it is axed to the required profile. For rubbed brick arches a soft brick called a rubber is used; the bricks are first cut to the approximate shape with a saw and are then finished off with an abrasive stone or file to produce the sharp arris. In both cases a template of plywood or hardboard to the required shape will be necessary for marking out the voussoirs. Typical examples of stone arches are shown in Fig. 3.1.4; the terminology and setting out of simple brick arches is shown in Fig. 3.4.1.

CENTRES

These are temporary structures, usually of light timber construction, which are strong enough to fulfil their function of supporting arches of brick or stone while they are being built and until they are sufficiently set to support themselves and the load over the opening. Centres can be an expensive item to a builder, therefore their design should be simple and adaptable so that as many uses as possible can be obtained from any one centre. A centre is always less in width than the soffit of an arch to allow for plumbing, that is, alignment and verticality of the face with a level or rule.

The type of centre to be used will depend upon:

- The weight to be supported.
- The span.
- The width of the soffit.

Generally soffits not wider than 150 mm will require one rib at least 50 mm wide and are usually called *turning pieces*. Soffits from 150 to 350 mm require two ribs which are framed together using horizontal tie members called *laggings*. Soffits over 350 mm require three or more sets of ribs. The laggings are used to tie the framed ribs together and to provide a base upon which the arch can be built. Close laggings are those which are touching each other, forming a complete seating for a gauged arch, and open laggings spaced at twice the width of the laggings, centre to centre, are used for rough arches.

If the arch is composed of different materials, for example, a stone arch with a relieving arch of brickwork, a separate centre for each material should be used. Typical examples of centres for brick arches are shown in Figs 3.4.2, 3.4.3 and 3.4.4.

Figure 3.4.2 Centres for small span arches

Outline of arch

200 × 25 rib

200 × 25 rib

25 × 25 laggings

200 × 25 rib

200 × 32 tie

200 × 25 tie

w.i. dogs

Braces

100 × 75 props

200 × 75 sole pieces

Ribs

Tie

Framed centre

Folding wedges

100 × 75 crosshead or headtree

100 × 32 braces

Prop

Sole piece

Figure 3.4.3 Typical framed centre for spans up to 1,500 mm

Framed centre

Folding wedges

100 × 100 head

Cross brace if required

200 × 32 horizontal brace

100 × 100 props

225 × 75 sole plate

Brace

200 × 32 horizontal brace

Outline of arch

Alt: open laggings at 50 c/c

Folding wedges

Head

250 × 32 braces

25 × 25 laggings

100 × 100 brace

Horizontal brace

Sole plate

250 × 32 tie

Props

Horizontal brace

250 × 32 ribs in 2 layers

250 × 32 tie

Figure 3.4.4 Typical framed centre for spans up to 4,000 mm

Figure 3.4.5 Arch form and cavity tray

CAVITY TRAYS

Arch construction can be simplified with structural support from a proprietory galvanized or stainless steel combined cavity tray and damp-proof course. Figure 3.4.5 shows the application and omission of conventional centring, saving considerable construction time as temporary supports are not required. The tray unifies the structural support with the damp-proof course and remains an integral part of the structure. Arch profiles are limited to tray manufacturers' designs and spans, but most will offer semi-circular and segmental as standard, with gothic, elliptical, ogee and triangular styles available to order. Clear spans up to about 2 m are possible.

TIMBER-FRAMED CONSTRUCTION

Timber framing of softwood members and sheet facings have a long association with non-loadbearing partition walls (see timber stud partitions – Chapter 7.7). Loadbearing timber walls for small buildings such as dwelling houses may be constructed by creating a framework of relatively small timbers (up to ex. 150 × 50 mm) spaced vertically at 400 or 600 mm. These are known as *studs* and have short struts of timber or noggins placed between them at 1 m maximum spacing to prevent distortion. Some panel prefabricators provide thin galvanized steel diagonal bracing as an alternative to struts. Head and sole plates complete the framing.

With the benefit of quality control procedures and efficient industrial manufacturing processes, the high cost of imported timber in the UK is offset by effective factory prefabrication of complete wall, floor and roof units. Furthermore, rationalized site assembly with mobile crane and semi-skilled operatives provides for considerable saving in construction time.

As a construction material, timber is comparatively light; it is therefore easy to handle and is less of a structural dead weight factor than traditional masonry. It has a high strength to weight ratio and is also very stiff when related to its strength. Prefabrication of structural units eliminates the need for skilled carpentry as all joints are simply nailed.

Apart from the cost of timber, the other disadvantages are poor resistance to fire in sizes under 150 × 100 mm, as there is insufficient substance to char protect the inner structure and its hygroscopic nature which can support decay in damp situations. The effect of fire is limited by cladding with non-combustible material such as plasterboard internally and facing brick externally. Provided the timber is installed dry and assembled with correct protection (dpc, vapour control layers, etc.), dampness penetration will not be an issue.

Manufacture and construction techniques derive from the applications of *platform* or *balloon* framing. Figure 3.5.1 shows the principle of storey-height platform-framed walls made up of ex. 100 × 50 mm timber studding at 400 or 600 mm centres. These are supported at intermediate floor levels or platforms by nailing through the sole plates. Head plates are similarly fixed to a head binder which links each prefabricated panel. Figure 3.5.2 shows the ground floor to eaves full-height balloon-framed wall. This is independent of the

Roof truss

Storey height wall unit

Intermediate floor

Ground floor

Eaves

ex. 100 × 50 head binder

ex. 100 × 50 head plate

Cavity barrier

ex. 100 × 50 studs at 400 or 600 centres

Breather membrane

Outer brick wall

8 mm wood fibre or plywood sheathing

Stud

Sole plate

50 mm header

19 mm plywood flooring

Floor joists at 400 mm centres

Head binder

Head plate

Intermediate floor

Cavity barrier

Stud

ex. 100 × 50 sole plate

ex. 100 × 50 sill plate

DPC

12 mm anchor bolts at 2 m spacing

Ground floor

Figure 3.5.1   Timber frame construction – platform frame

Figure 3.5.2　Timber frame construction – balloon frame

Note: Eaves and ground floor as Fig. 3.5.1

intermediate floor except for complementary bracing where the floor joists fix to studs and a *ribbon* or *ledger* let into the stud framing. Storey-height platform framing is simpler to handle and transport but less rapidly constructed on site than the one lift to roof level that balloon framing permits. The effects of timber movement are more easily contained and restrained in the smaller units of a platform frame.

An outer cladding of brickwork provides a traditional appearance, with galvanized steel angle wall ties spanning a cavity to secure the masonry outer leaf to the timber framing. The possibility of fire spread through the cavity is restrained with cavity barriers at strategic intervals. These normally occur at intersections of elements of construction, e.g. wall to roof and party wall to external wall. Fire resistance of cavity barriers in dwellings should be at least 30 minutes, i.e. 30 minutes integrity and 15 minutes insulation (see Building Regulations: Approved Document B3: Section 9, BS 476: *Fire tests on building materials and structures*, and Fig. 3.5.3).

12.5 mm plasterboard

Breather membrane

Alternative

500 gauge polyethylene vapour control membrane

80/100 mm mineral wool insulation between framing members

8 mm min. plywood sheathing

Brickwork outer leaf with wall ties every 5th. course

Foil-faced glass fibre reinforced polyisocyanurate (Pi) foam board (10–67 mm) [Cellular plastic insulation board]

External wall construction

90 mm cavity

Mineral wool sound insulation quilt flanged for stapling to framing (preferably in both stud frames)

19 mm plasterboard (horizontal)

12.5 mm plasterboard (vertical)

50 mm wire meshed mineral wool fire/cavity barrier

Floor boarding

Floor joist

50 mm timber blocking between joists

Restraint ties at 1.2 m spacing

Party wall construction

Figure 3.5.3 Timber frame construction – further details

TIMBER: PROPERTIES AND GRADING

Timber derives from thousands of different tree species, but the botanical classification divides timber into *hardwood* and *softwood*. Hardwoods are from broad-leaved trees, most of which are deciduous, holly being an exception. Not all hardwoods are hard; balsa is an example of a *soft* hardwood. Softwoods are from coniferous trees and not all of these are relatively soft, e.g. yew is a very strong, dense and durable material. For structural purposes, strength characteristics and durability are more important means for classification. This is particularly important with softwood which is used predominantly for structural timbers in the UK, due to the prohibitive cost of hardwood.

When specifying softwood structural timbers, the following should be stated:

- Any preservative treatment
- Cross-sectional size and surface finish
- The moisture content
- The strength class

PRESERVATIVE TREATMENT

The majority of softwood contains a high proportion of sapwood. This is less resilient than the denser heartwood found closer to the core of a timber trunk. Therefore, as a precaution against dampness and possible fungal decay (wet or dry rot), plus potential infestation from wood-boring insects, suitable preservatives may be required. The need will be determined by the position of timbers within a building and the geographical location. In parts of southern England there is a perceived risk of softwood timber infestation by the *house longhorn beetle*. The Building Regulations, Approved Document to support Regulation 7, defines the particular boroughs where softwoods must be treated if applied in any aspect of roof construction. See also BS 5268: Part 5: *Code of practice for the preservative treatment of structural timber*, for clarification of acceptable chemical treatment.

CROSS-SECTIONAL SIZE AND SURFACE FINISH

Softwood cross-sectional size has previously been specified under three classifications: sawn, planed (machined) and regularized. Sawn is the result of converting a timber log into commercially accepted sizes, e.g. 50 mm × 100 mm, 75 mm × 225 mm. Application of these sections is largely for unseen structural use, such as floor and roof members. Planed involves machining about 1.5 mm off each surface to provide a smooth finish, e.g. 50 mm × 100 mm sawn becomes 47 mm × 97 mm, otherwise known as ex. 50 × 100. The 'ex' meaning 'out of'. Regularizing has been used to achieve a uniform width in timber to provide a more even final product. The timber may be sawn or planed to, say, 47 mm for uniformity in timber studding.

However, these terms are now redefined, with regularizing no longer featured in the requirements of BS EN 336: *Structural timber – Coniferous and poplar – Sizes – Permissible deviations*. This standard specifies *target* sizes in order to simplify the previous confusion over the finished dimensions. Target size is in only two tolerance classes: T1, applicable to sawn surfaces, and T2, applicable to planed timber. For example, if a section of timber is required 50 mm wide sawn and 197 mm in depth planed, it will be specified as 50 mm (T1) × 197 mm (T2).

MOISTURE CONTENT

When timber is felled it is saturated. In this state it is unworkable and, if allowed to remain so, will be prone to shrinkage, distortion and the effects of fungal growths. Natural drying or seasoning is possible in some climates, but most commercial organizations use kilns to reduce the moisture content of rough sawn softwood to between 12 and 20%. This is expressed as the weight of water in the timber as a percentage of the weight of the timber dry, i.e.:

$$\text{moisture content } \% = \frac{(\text{wet weight}) - (\text{dry weight})}{\text{dry weight}} \times \frac{100}{1}$$

After seasoning it is essential that the product remains in a well ventilated atmospherically stable environment. Timber is hygroscopic, i.e. made up of cells which will absorb moisture readily. Therefore significant changes in temperature and humidity may cause it to react by expanding, contracting, deforming or twisting. Moisture content of stored timber can be checked regularly with a surveyor's moisture meter. These have two pointed probes and respond to electrical resistance (moisture) through the material.

STRENGTH CLASSES

Grading of timber may be visual or by the more efficient use of computerized grading machines. Individual pieces are assessed against permissible defects limits and marked accordingly. Some examples of grade markings are shown in Fig. 3.6.1.

Figure 3.6.1 Examples of grading stamps on softwood timber

Grading occurs in this country or the country of origin. BS EN 518 and 519 (European standards for visual and machine grading, respectively) effectively rationalizes much of the timber graded in Europe and many surrounding countries, e.g. Scandinavia, but confusion may occur with timber graded elsewhere in the absence of an international standard. Visual grading in the UK may be in accordance with BS 4978. This provides two quality standards, GS (general structural) and SS (special structural). The letter M preceding GS or SS indicates grading by machine. GS and SS grades will meet the requirements of the aforementioned European standard.

Timber imported from North America is independently graded to rules applicable in Canada and the USA. The Canadian National Lumber Grades Authority (NLGA) and the American National Grading Rules for Dimension Lumber (NGRDL) grades in accordance with product application, see Table 3.6.1.

Comparisons can be drawn and Table 3.6.2 provides an indication of the similarity of product grades and strength classes to BS EN 338: *Structural timber – strength classes.*

Table 3.6.1   North American timber grades and markings

Timber mark	Grade	Application	Range of nominal widths
Sel str No. 1 No. 2 No. 3	Select structural Number 1 Number 2 Number 3	Structural joists and planks	150 mm and over
Sel str No. 1 No. 2 No. 3	Select structural Number 1 Number 2 Number 3	Structural light framing	50 mm to 100 mm
Const Std Util Stud	Construction Standard Utility Stud	Light framing	50 mm to 100 mm

Table 3.6.2   Timber grading and strength class comparisons

Species	Standard	C14	C16	C18	C22	C24	C27/TR26	C30
	BS EN 338							
	Bldg Regs		SC3			SC4	SC5	
Whitewood or redwood	BS 4978		GS			SS		
British spruce	BS 4978	GS		SS				
British pine	BS 4978	GS			SS			
Canadian S-P-F or hem-fir	NLGA or NGRLD (J&P/SLF)		No. 1 No. 2			Sel		
	NLGA or NGRLD (LF)	Const						
	NLGA or NGRLD (Stud)	Stud						

Key: S-P-F = Spruce, pine, fir.
hem   = Hemlock.
Others, see Table 3.6.1.

Notes
1. North American machine stress rated (MSR) grades are also accepted in the UK and correspond across the BS EN 338 range given above.
2. For a full and more detailed comparison, refer to BS 5268: Part 2: *Structural use of timber*.

As can be seen from Table 3.6.2, using GS or SS graded timber from different sources may not result in the same strength characteristics, as these can vary between species. The full listing of BS EN 338 strength classes are shown in Table 3.6.3, extracted from BS 5268: Part 2: 1996 (Table 7) with kind permission

Table 3.6.3 Grade stresses and moduli of elasticity for various strength classes: for service classes 1 and 2

Strength class	Bending parallel to grain (N/mm²)	Tension parallel to grain (N/mm²)	Compression parallel to grain (N/mm²)	Compression perpendicular to grain* (N/mm²)	Shear parallel to grain (N/mm²)	Modulus of elasticity		Characteristic density, ρ_k† (kg/m³)	Average density, ρ_{mean}† (kg/m³)
						Mean (N/mm²)	Minimum (N/mm²)		
C14	4.1	2.5	5.2	2.1	1.6	0.60	6,800	4,600	290	350
C16	5.3	3.2	6.8	2.2	1.7	0.67	8,800	5,800	310	370
C18	5.8	3.5	7.1	2.2	1.7	0.67	9,100	6,000	320	380
C22	6.8	4.1	7.5	2.3	1.7	0.71	9,700	6,500	340	410
C24	7.5	4.5	7.9	2.4	1.9	0.71	10,800	7,200	350	420
TR26‡	10.0	6.0	8.2	2.5	2.0	1.10	11,000	7,400	370	450
C27	10.0	6.0	8.2	2.5	2.0	1.10	12,300	8,200	370	450
C30	11.0	6.6	8.6	2.7	2.2	1.20	12,300	8,200	380	460
C35	12.0	7.2	8.7	2.9	2.4	1.30	13,400	9,000	400	480
C40	13.0	7.8	8.7	3.0	2.6	1.40	14,500	10,000	420	500
D30	9.0	5.4	8.1	2.8	2.2	1.40	9,500	6,000	530	640
D35	11.0	6.6	8.6	3.4	2.6	1.70	10,000	6,500	560	670
D40	12.5	7.5	12.6	3.9	3.0	2.00	10,800	7,500	590	700
D50	16.0	9.6	15.2	4.5	3.5	2.20	15,000	12,600	650	780
D60	18.0	10.8	18.0	5.2	4.0	2.40	18,500	15,600	700	840
D70	23.0	13.8	23.0	6.0	4.6	2.60	21,000	18,000	900	1,080

Note: Strength classes C14 to C40 and TR 26 are for softwoods, D30 to D70 are for hardwoods.
* When the specification specifically prohibits wane at bearing areas, the higher values of compression perpendicular to grain stress may be used, otherwise the lower values apply.
† The values of characteristic density given above are for use when designing joints. For the calculation of dead load, the average density should be used.
‡ The strength class TR26 is essentially for the manufacture of trussed rafters but may be used for other applications with the grade stresses and moduli given above. When used with the provisos given in BS 5268: Part 3 the tabulated permissible loads for strength class C27 should be used. For joints, the grade stresses are similar to the former M75 redwood/whitewood so timber and trussed rafter designs to this M75 grade/species combination are interchangeable with timber and trussed rafter designs using the TR26 strength class.

of The British Standards Institution. It provides a comprehensive comparison of timber strength properties and characteristics, which may be used as a basis for structural design and timber selection.

Note: Service classes 1 and 2 in Table 3.6.3 refer to timber used in moderate climates where ambient temperature is about 20 °C and relative humidity does not exceed 65% and 85% respectively for more than a few weeks per year. Respective moisture contents will not exceed 12% and 20%. Service class 3 exists where climatic conditions impose higher moisture contents than service class 3.

PART 4

FLOORS

FLOORING: SOLID GROUND, SUSPENDED TIMBER

SOLID GROUND FLOORS

The construction of a solid ground floor can be considered under three headings:

- Hardcore.
- Blinding.
- Concrete bed or slab.

HARDCORE

The purpose of hardcore is to fill in any small pockets which have formed during oversite excavations, to provide a firm base on which to place a concrete bed and to help spread any point loads over a greater area. It also acts against capillary action of moisture within the soil. Hardcore is usually laid in 100–150 mm layers to the required depth and it is important that each layer is well compacted, using a roller if necessary, to prevent any unacceptable settlement beneath the solid floor.

Approved Document C recommends that no hardcore laid under a solid ground floor shall contain water-soluble sulphates or other harmful matter in such quantities as to be liable to cause damage to any part of the floor. This recommendation prevents the use of any material which may swell upon becoming moist, such as colliery shale, and furthermore it is necessary to ascertain that brick rubble from demolition works and clinker furnace waste intended for use as hardcore does not have any harmful water-soluble sulphate content.

BLINDING

This is used to even off the surface of hardcore if a damp-proof membrane is to be placed under the concrete bed or if a reinforced concrete bed is specified. Firstly, it will prevent the damp-proof membrane from being punctured by the hardcore and, secondly, it will provide a true surface from which the reinforcement can be positioned. Blinding generally consists of a layer of sand 25–50 mm thick or a 50–75 mm layer of weak concrete (1:12 mix usually suitable) if a true surface for reinforced concrete is required.

CONCRETE BED

Thicknesses generally specified are:

- Unreinforced or plain *in-situ* concrete, 100–150 mm thick.
- Reinforced concrete, 150 mm minimum.

Suitable concrete mixes are:

- Plain *in-situ* concrete – 50 kg cement:0.11 m^3 fine aggregate:0.16 m^3 coarse aggregate or mix specification ST2 to BS 5328.
- Reinforced concrete – 50 kg cement:0.08 m^3 fine aggregate:0.13 m^3 coarse aggregate or mix specification ST4 to BS 5328.

The reinforcement used in concrete beds for domestic work is usually in the form of a welded steel fabric to BS 4483. Sometimes a light square mesh fabric is placed 25 mm from the upper surface of the concrete bed to prevent surface crazing and limit the size of any cracking.

In domestic work the areas of concrete are defined by the room sizes and it is not usually necessary to include expansion or contraction joints in the construction of the bed.

Protection of floors next to the ground

Building Regulation C2 requires that such part of a building as is next to the ground shall have a floor so constructed as to prevent the passage of moisture from the ground to the upper surface of the floor. The requirements of this regulation can only be properly satisfied by the provision of a suitable barrier in the form of a damp-proof membrane within the floor. The membrane should be turned up at the edges to meet and blend with the damp-proof course in the walls to prevent any penetration of moisture by capillary action at edges of the bed.

Suitable materials for damp-proof membranes are:

- Polythene (LDPE) 1,200 gauge (0.3 mm) sheet with sealed joints is acceptable and will also give protection against moisture vapour as well as moisture.
- Hot poured bitumen – should be at least 3 mm thick.
- Cold applied bitumen/rubber emulsions – should be applied in not less than three coats.
- Asphalt/pitchmastic – could be dual-purpose finish and damp-proof membrane.

Note: Prevention of the ingress of radon and/or methane gases from the ground will require a wire or fibre reinforced LDPE membrane of up to 1 mm thickness. Radon is a naturally occurring radioactive gas originating from uranium and radium deposits in certain rock subsoils found primarily in parts of the west country, northern England and areas of Scotland. Methane is an explosive gas which can build up in the ground as a result of deposited decaying organic materials. Where either of these gases is prevalent, it is now standard practice to construct suspended precast concrete floors with natural draught underfloor ventilation – see Chapter 4.3.

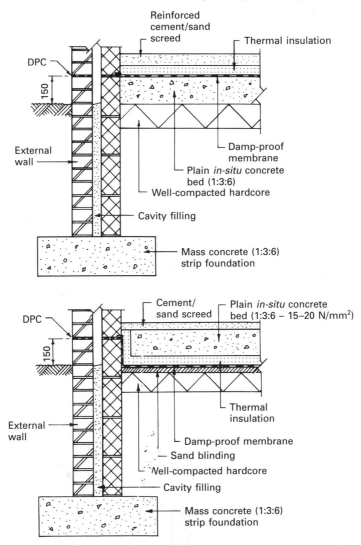

Figure 4.1.1 Typical solid floor details at external walls

The position of a damp-proof membrane, whether above or below the concrete bed, is a matter of individual choice. A membrane placed above the bed is the easiest method from a practical aspect and is therefore generally used. A membrane placed below the bed has two advantages: firstly, it will keep the concrete bed dry and in so doing will make the bed a better thermal insulator and, secondly, during construction it will act as a separating layer preventing leakage of the cement matrix into the hardcore layer which could result in a weak concrete mix. Typical details of solid floor construction are shown in Figs 4.1.1 and 4.1.2.

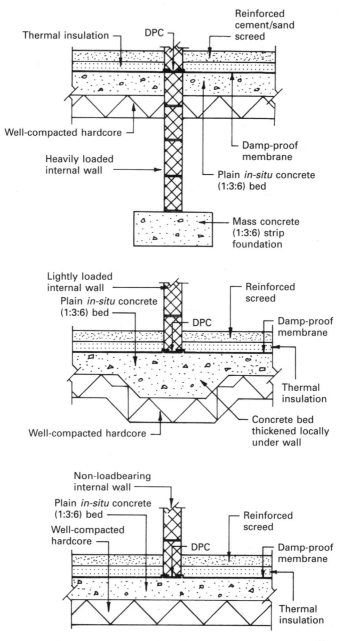

Figure 4.1.2   Typical solid floor details at internal walls

SUSPENDED TIMBER GROUND FLOORS

This type of floor consists of timber boards or other suitable sheet material fixed to joists spanning over sleeper walls and was until 1939 the common method of forming ground floors in domestic buildings. The Second World War restricted the availability of suitable timber and solid ground floors replaced suspended timber floors. Suspended precast concrete floors have gained popularity in recent times, not only as a precaution against radon and methane, but as a less expensive form of construction. These are considered in more detail in Chapter 4.3. Today the timber floor is still used on occasions because it has some flexibility and will easily accept nail fixings – properties which a solid ground floor lacks. It is a more expensive form of construction than a concrete floor and can only be justified to match existing construction where a building is extended. It could be used on sloping sites which require a great deal of filling to make up the ground to the specified floor level, although a precast concrete flooring system could also be used, and this is likely to be much cheaper.

Suspended timber ground floors are susceptible to dry rot, draughts and are said to be colder than other forms of flooring. If the floor is correctly designed and constructed these faults can be eliminated.

The problem of dry rot, which is a fungus that attacks damp timber, can be overcome by adequate ventilation under the floor and the correct positioning of damp-proof courses to keep the under floor area and timber dry. Through ventilation is essential to keep the moisture content of the timber below that which would allow fungal growth to take place; that is, 20% of its oven-dry weight. The usual method is to allow a free flow of air under the floor covering by providing air bricks in the external walls. These are sited near the corners and at approximately 2 m centres around the perimeter of the building. They must have an equivalent of 1,500 mm^2/m run of wall or alternatively 500 mm^2/m^2 of floor area will also be acceptable (take greater value). If a suspended timber floor is used with an adjacent solid ground floor, pipes of 100 mm diameter are used under the solid floor to convey air to and from the external walls to the suspended floor.

Building Regulations

Building Regulation C2 applies as with solid floors and recommended provisions are given in Approved Document C. Figure 4.1.3 shows the minimum dimensions recommended in Approved Document C but in practice a greater space between the concrete bed and the timber is usual. The honeycomb sleeper walls are usually built two or three courses high to allow good through ventilation. Sleeper walls spaced at 2,000 mm centres will give an economic joist size. The width of joists is usually taken as 50 mm, this will give sufficient width for the nails securing the covering, and the depth can be obtained by reference to design tables recommended in Approved Document A or by design calculations. The usual joist depth for domestic work is 125 mm.

Cement particle board connecting piece

Air brick of galv. cast iron, plastic or terra cotta

Typical air brick detail

Thermal insulation between joists

t & g boards

Joists at 400 c/c fixed clear of external wall

150 min.

75 min.

Honeycomb built sleeper wall

DPC below level of wall plate

Well-compacted hardcore

Mass concrete strip foundation

Underfloor space to be clear of debris and have through ventilation

Concrete not less than 100 thick composed of 50 kg cement:0.13 m³ fine aggregate:0.18 m³ coarse aggregate laid entirely above highest adjoining ground unless laid to falls to drainage outlet at lowest level of adjoining ground

Air bricks connected with slate trunking or similar

Cavity to extend 150 below lowest DPC

150

Cavity filling

Figure 4.1.3 Building regulations and suspended timber floors

Layout

The most economic layout is to span the joists across the shortest distance of the room, this means that joists could be either parallel or at right angles to a fireplace. The fireplace must be constructed of non-combustible materials and comply with Building Regulation J3. Typical examples are shown is Figs 4.1.4 and 4.1.5.

Insulation in ground floors

Rigid polystyrene or high-density mineral wool can be incorporated within a solid floor structure to comply with the Building Regulations Approved Document L – Conservation of Fuel and Power. Location may be above the damp-proof membrane and below the concrete slab or between the screed and concrete slab. If the latter, the screed should be at least 75 mm thickness and wire reinforced against cracking. Suspended timber floors can incorporate mineral fibre insulation draped on nylon netting between the joists, or have rigid insulation boards spanning the gap between joists. More detail of thermal insulation in floors is provided in Chapter 8.2.

SUSPENDED TIMBER UPPER FLOORS

Timber, being a combustible material, is restricted by Part B of the Building Regulations to small domestic buildings as a structural flooring material. Its popularity in this context is due to its low cost in relationship to other structural flooring methods and materials. Structural softwood is readily available at a reasonable cost, is easily worked and has a good strength to weight ratio and is therefore suitable for domestic loadings.

Terminology

Common joist   A joist spanning from support to support.

Trimming joist   Span as for common joist but it is usually 25 mm thicker and supports a trimmer joist.

Trimmer joist   A joist at right angles to the main span supporting the trimmed joists and is usually 25 mm thicker than a common joist.

Trimmed joist   A joist cut short to form an opening and is supported by a trimmer joist; it spans in the same direction as common joists and is of the same section size.

125 mm thick concrete
constructional hearth

Rubble-filled void

DPC

External wall

T & G flooring nailed to joists

328

500

150

Half brick
fender wall

125 × 50 joists at 400 c/c

100 × 75
wall plate

32 wide strap
fixings at
750 c/c

DPC

Half-brick honeycomb
sleeper walls at
2.000 c/c

Mass concrete bed

Well-compacted hardcore

Mass concrete strip foundation

Note: Detail historic, therefore blockwork and insulation omitted

Figure 4.1.4 Typical details of suspended floor – joists parallel to fireplace

125 mm thick concrete constructional hearth

Rubble filled void

DPC

External wall

Sleeper wall

125 × 50 joists at 400 c/c

T & G flooring nailed to joists

100 × 75 wall plate

32 wide strap fixings at 750 c/c

DPC

Half brick honeycomb sleeper walls at 2.000 c/c

Mass concrete bed

Well compacted hardcore

Mass concrete strip foundation

Fender wall

328

500

150

Note: Detail historic, therefore blockwork and insulation omitted

Figure 4.1.5 Typical details of suspended timber floor – joists at right angles to fireplace

JOIST SIZING

There are three ways of selecting a suitable joist size for supporting a domestic type floor:

1. Rule of thumb:

$$\frac{\text{span in mm}}{24} + 50 \text{ mm} = \text{depth in mm}.$$

2. Calculation:

$$\text{BM} = \frac{fbd^2}{6}$$

   where   BM = bending moment
   f = maximum fibre stress (see Chapter 3.6)
   b = breadth (assumed to be 50 mm)
   d = depth in mm.
3. Building Regulations, Approved Document A: "Timber Intermediate Floors for Dwellings", published by TRADA.

Joists

If the floor is framed with structural softwood joists of a size not less than that required by the Approved Document, the usual width is taken as 50 mm. The joists are spaced at 375–450 mm centre to centre depending on the width of the ceiling boards which are to be fixed on the underside. Maximum economy of joist size is obtained by spanning in the direction of the shortest distance to keep within the deflection limitations allowed. The maximum economic span for joists is between 3,500 and 4,500 mm; for spans over this a double floor could be used.

Support

The ends of the joists must be supported by loadbearing walls. The common methods are to build in the ends or to use special metal fixings called *joist hangers*; other methods are possible but these are seldom employed. Support on internal loadbearing walls can be by joist hangers or direct bearing when the joists are generally lapped (see Fig. 4.1.6).

Trimming

This is a term used to describe the framing of joists around an opening or projection. Various joints can be used to connect the members together, all of which can be substituted by joist hangers. Trimming around flues and upper floor

Typical joist hangers

Note: Insulation in external wall cavity if required (see Ch. 8.2)

Figure 4.1.6 Typical joist support details

fireplaces should comply with the recommendations of Approved Document J. It should be noted that, since central heating is becoming commonplace, the provision of upper floor fireplaces is seldom included in modern designs, because they are considered to be superfluous. However, they are frequently encountered in renovation and maintenance work to older properties. Typical trimming joints and arrangements are shown in Figs 4.1.7 and 4.1.8.

Housed joint   **Bevelled housed joint**

Tusked tenon joint

Typical stairwell trimming

Figure 4.1.7   Floor trimming joints and details

Trimming to upper floor fireplace

Trimming to stairwell

Trimming around flues

Note: Recess constructed of solid non-conbustible material.

Figure 4.1.8 Typical trimming arrangements

Strutting

Shrinkage in timber joists will cause twisting to occur and this will result in movement of the ceiling below and could cause the finishes to crack. To prevent this strutting is used between the joists if the total span exceeds 2,400 mm; the strutting being placed at mid–span (see Fig. 4.1.9).

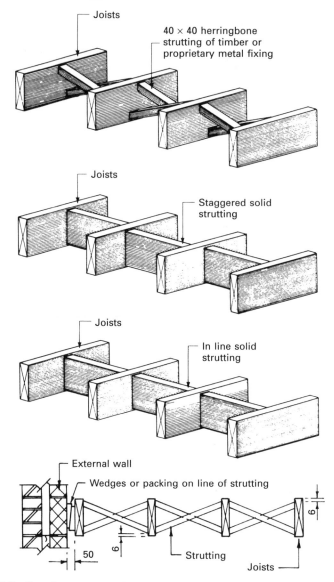

Figure 4.1.9 Strutting arrangements

Double floors

These can be used on spans over 4,500 mm to give a lower floor area free of internal walls. They consist of a steel beam or timber binder spanning the shortest distance which supports common joists spanning at right angles. The beam reduces the span of the common joists to a distance which is less than the shortest span to allow an economic joist section to be used. The use of a timber binder was a

Figure 4.1.10   Double floor details

popular method but it is generally considered to be uneconomic when compared with a standard steel beam section. Typical details are shown in Fig. 4.1.10.

If the span is such that a double floor is deemed necessary it would be a useful exercise to compare the cost with that of other flooring methods, such as *in-situ* reinforced concrete and precast concrete systems, which, overall, could be a cheaper and more practical solution to the problem.

FLOORING: RAISED ACCESS FLOORS

These are a system of elevated platforms suspended over the structural floor of *in-situ* or precast concrete. They are not new in concept and form the basis of deck construction popular about the outside of North American homes. Used internally, they have evolved primarily in response to the information technology revolution occurring in the latter part of the twentieth century. While the huge volume of power, telephone and data cabling in modern offices has been largely responsible for developments in raised platforms, location for service pipes and ventilation ducts has also been influential. The dominance of services issues has promoted a re-think for designers, with the need for greater floor to floor heights to incorporate these floor voids (and suspended ceilings).

PARTIAL ACCESS

This is where the floor finish is secured and access is limited, possibly into simple ducts or trunking as shown in Fig. 4.2.1. They are most suited for access to mains cabling for power and lighting, with void depths limited to about 100 mm. Accesses should be positioned to avoid furniture and be strategically located over junction boxes. This provides flexibility for changes in cable distribution where work stations and work functions may change.

FULL ACCESS

These comprise standard size (600 mm × 600 mm) interchangeable floor panels elevated on adjustable pedestals to provide multi-directional void space. Heights usually vary between 100 and 600 mm, but extremes of as little as 50 mm and as much as 2 m are possible. Adjustable height pedestals are manufactured from steel or polypropylene, with support plates containing four lugs or projections to locate loose fit decking panels. Threaded flat support plates may be specified where panels are to be screw fixed in place (see Fig. 4.2.2).

Floor panels or decking materials can vary considerably to suit application. Timber joists or battens may be used to support softwood boards or plywood/

172

Figure 4.2.1 Floor duct

Figure 4.2.2 Adjustable pedestal

Figure 4.2.3   Raised access floor

chipboard sheets, particularly if matching a traditional suspended floor. Office specification usually requires a fully interchangeable system with standard 600 mm × 600 mm panels. These are available in various gradings including light, medium and heavy, with corresponding maximum loadings of 10, 20 and 50 kN/m² . Panel construction and application is shown in Fig. 4.2.3.

The combination of fully bonded chipboard within a steel casing provides structural soundness and integrity in the event of fire. The chipboard acts as a fire, acoustic and thermal insulator, and the steel resists tensile loading, also protecting the core against spread of flame. Presence of extraneous metal requires contact protection by electrical earth continuity through the complete floor system.

CHAPTER 4.3

PRECAST CONCRETE FLOORS

The function of any floor is to provide a level surface which is capable of supporting all the live and dead loads imposed. Reinforced concrete with its flexibility in design, good fire resistance and sound insulating properties is widely used for the construction of suspended floors for all types of buildings. The disadvantages of *in-situ* concrete are:

- Need for formwork.
- Time taken for the concrete to cure before the formwork can be released for reuse and the floor available as a working area.
- Very little is contributed by a large proportion of the concrete to the strength of the floor.

Floors composed of reinforced precast concrete units have been developed over the years to overcome some or all of the disadvantages of *in-situ* reinforced concrete slab. To realize the full economy of any one particular precast flooring system the design of the floors should be within the span, width, loading and layout limitations of the units under consideration; coupled with the advantages of repetition.

CHOICE OF SYSTEM

Before any system of precast concrete flooring can be considered in detail the following factors must be taken into account:

- Maximum span.
- Nature of support.
- Weight of units.
- Thickness of units.
- Thermal insulation properties.
- Sound insulation properties.
- Fire resistance of units.

- Speed of construction.
- Amount of temporary support required.

The systems available can be considered as either precast hollow floors or composite floors; further subdivision is possible by taking into account the amount of temporary support required during the construction period.

PRECAST HOLLOW FLOORS

Precast hollow floor units are available in a variety of sections such as box planks or beams; tee sections, I beam sections and channel sections (see Fig. 4.3.1). The economies which can be reasonably expected over the *in-situ* floor are:

- 50% reduction in the volume of concrete.
- 25% reduction in the weight of reinforcement.
- 10% reduction in size of foundations.

The units are cast in precision moulds, around inflatable formers or foamed plastic cores. The units are laid side by side with the edge joints being grouted together; a structural topping is not required but the upper surface of the units are usually screeded to provide the correct surface for the applied finishes (see Fig. 4.3.1). Little or no propping is required during the construction period but usually some means of mechanical lifting is required to offload and position the units. Hollow units are normally the cheapest form of precast concrete suspended floor for simple straight spans with beam or wall supports up to a maximum span of 20.000 m. They are not considered suitable where heavy point loads are encountered unless a structural topping is used to spread the load over a suitable area.

The hollow beams or planks give a flat soffit which can be left in its natural state or be given a skim coat of plaster; the voids in the units can be used to house the services which are normally incorporated in the depth of the floor. The ribbed soffit of the channel and tee units can be masked by a suspended ceiling; again the voids created can be utilized to house the services. Special units are available with fixing inserts for suspended ceilings, service outlets and edges to openings.

COMPOSITE FLOORS

These floors are a combination of precast units and *in-situ* concrete. The precast units, which are usually prestressed or reinforced with high yield steel bars, are used to provide the strength of the floor with the smallest depth practicable and at the same time act as permanent formwork to the *in-situ* topping which provides

Spans up to 13.000

Typical hollow floor unit details

Anti-crack reinforcement required if units
are continuous over internal supports

Typical channel section floor unit details

Figure 4.3.1   Precast concrete hollow floors

the compressive strength required. It is essential that an adequate bond is achieved
between the two components – in most cases this is provided by the upper surface
texture of the precast units; alternatively a mild steel fabric can be fixed over the
units before the *in-situ* topping is laid.

Composite floors generally take one of two forms:

Figure 4.3.2   Composite floors – prestressed plank

1. Thin prestressed planks with a side key and covered with an *in-situ* topping (see Fig. 4.3.2).
2. Reinforced or prestressed narrow beams which are placed at 600 mm centres and are bridged by concrete filler blocks known as *pots*. The whole combination is covered with *in-situ* structural concrete topping. Most of the beams used in this method have a shear reinforcing cage projecting from the precast beam section (see Fig. 4.3.3).

In both forms temporary support should be given to the precast units by props at 1.800 to 2.400 m centres until the *in-situ* topping has cured.

COMPARISON OF SYSTEMS

Precast hollow floors are generally cheaper than composite, *in-situ* concrete is not required and therefore the need for mixing plant and storage of materials is eliminated. The units are self-centring, therefore temporary support is not required, the construction period is considerably shorter and generally the overall weight is less.

Composite floors will act in the same manner as an *in-situ* floor and can therefore be designed for more complex loadings. The formation of cantilevers is easier with this system and support beams can be designed within the depth of the floor giving a flat soffit. Services can be housed within the structural *in-situ* topping, or within the voids of the filler blocks. Like the precast hollow floor, composite floors are generally cheaper than a comparable *in-situ* floor within the limitations of the system employed.

Typical composite floor using PCC planks

Typical composite floor using PCC beams

Figure 4.3.3 Composite floors – beam and pot

BEAM AND BLOCK FLOOR

This combines the principles of the precast hollow and composite floor systems. It has developed into a cost- and time-effective means of constructing domestic ground and upper floors, by incorporating precast concrete beams with lightweight

Figure 4.3.4 Beam and block domestic ground floor

concrete blocks as an in filling. The benefits of quality controlled factory manufacture of components and simple site assembly with the aid of a mobile crane to hoist the beams, add to the following advantages:

1. Potential to span over unsound infilling, common to sloping sites.
2. Application over movable subsoils such as shrinkable clay (see Chapter 2.7).
3. Suitability where ventilation under the ground floor is required to dilute intrusive gases.

The Building Regulations, Approved Document C, requires a minimum clear void depth of 75 mm below these floors, but it is usual practice to leave at least 150 mm. Ventilation of the void is advisable to dilute and prevent concentration of gases from the ground (radon and/or methane) and possible leakage from piped services. Figure 4.3.4 shows typical construction of a domestic suspended beam and block floor, where the stripped topsoil leaves the underfloor surface lower than adjacent ground. This is only acceptable if the soil is free draining. Also, ground differentials should be minimal, otherwise the external wall becomes a retaining wall and will require specific design calculations. All organic material should be removed from the void and the surface treated with weedkiller. Void depth may need to be as much as 225 mm in the presence of heavy clay subsoil and nearby trees (see Chapter 2.7).

Upper floors follow the same principles of assembly, with purpose-made trimmer shoes providing support to concrete beams around stair openings. Span potential is only about 5 m, therefore intermediate support from a loadbearing partition or a steel beam is acceptable. Direct flange bearing is possible, but if a deep section is required the floor structure may be accommodated on ledger angles. Typical upper floor details are shown in Fig. 4.3.5.

Figure 4.3.5 Beam and block – upper floor intermediate support

HOLLOW BLOCK AND WAFFLE FLOORS

Precast concrete suspended floors are generally considered to be for light to medium loadings spanning in one direction. Hollow block, or hollow pot floors as they are sometimes called, and waffle or honeycomb floors can be used as an alternative to the single spanning precast floor since they can be designed to carry heavier loadings. They are in fact ribbed floors consisting of closely spaced narrow and shallow beams giving an overall reduction in depth of the conventional reinforced concrete *in-situ* beam and slab floor.

HOLLOW BLOCK FLOORS

These are formed by laying over conventional floor soffit formwork a series of hollow lightweight clay blocks or pots in parallel rows with a space between these rows to form the ribs. The blocks act as permanent formwork giving a flat soffit suitable for plaster application and impart to the floor good thermal insulation and fire resistance. The ribs formed between the blocks can be reinforced to suit the loading conditions of the floor, thus providing flexibility of design (see Fig. 4.4.1). The main advantages of this system are its light weight, which is generally less than comparable floors of concrete construction, and its relatively low cost.

WAFFLE OR HONEYCOMB FLOORS

Used mainly as an alternative to an *in-situ* flat slab or a beam and slab suspended floor since it requires less concrete, less reinforcement and can be used to reduce the number of beams and columns required with the resultant savings on foundations. The honeycomb pattern on the underside can add to the visual aspect of the ceiling by casting attractive shadow patterns.

The floor is cast over lightweight moulds or pans made of glass fibre, polypropylene or steel forming a two-directional ribbed floor (see Fig. 4.4.2). The moulds are very strong, lightweight and are capable of supporting all the normal loads encountered in building works. Support is reduced to the minimum since the moulds are arranged in parallel rows and span between the parallel lines of temporary supports.

182

Projecting top reinforcement

RC edge beam

In-situ concrete topping

Hollow clay floor blocks

Reinforcement to design

In-situ concrete ribs

Ceiling finish

Filler tiles (optional)

75, 100 or 125

Voids

295

295

Rib reinforcement

Voids in end blocks closed with void filler tiles or any suitable material

Filler tiles

Figure 4.4.1 Hollow block floors

Typical waffle floor mould

Plan size 800 × 800
depths 225, 300 & 375

Rib reinforcement to design

In-situ ribs at 800 c/c

In-situ ribs at 800 c/c

50 mm thick *in-situ* topping reinforced with welded fabric

RC edge beam

Honeycomb indents left by moulds

RC edge beam

Figure 4.4.2 Waffle or honeycomb floors

The reinforcement in the ribs is laid in two directions to resist both longitudinal and transverse bending moments in the slab. Generally three mould depths are available but the overall depth can be increased by adding to the depth of the topping.

With all floors using an *in-situ* topping it is possible to float the surface in preparation for the applied finishes, but this surface may suffer damage while being used by the following trades. It is therefore considered a better form of construction to allow for a floor screed to be applied to the *in-situ* topping at a later stage in the contract prior to the fixing of the applied finish.

LATERAL RESTRAINT AND SLENDERNESS FACTORS

Walls have been designed such that their mass, together with the loads carried from floors and roofs, provide adequate resistance to buckling. Buckling can be controlled by using buttresses or piers, but these may encroach on space and be visually unattractive. Contemporary design favours more slender construction, therefore stability and resistance to tensile lateral forces such as wind and possibly impact, is achieved by using support from the floor and roof (see Part 6).

Floor structures function as a stable horizontal platform supported by perimeter and intermediate walls. They complement that support by providing restraint and stability for the wall against lateral overturning forces, in conjunction with adequate structural connections. Connections to external, compartment (fire), separating (party), and internal loadbearing walls in houses where walls exceed 3 m length may be achieved by:

1. 90 mm minimum direct bearing of the floor by timber joists (see Fig. 4.1.6), or from a continuous concrete slab.
2. BS 5628: Part 1 approved galvanized steel restraint type joist hangers with a minimum 100 mm bearing, spaced at no more than 2 m.
3. Galvanized steel purpose-made straps of minimum cross-section 30×5 mm, spaced at no more than 2 m, secured to the inner leaf of masonry and over at least three joists.
4. Galvanized steel straps, as 3 above, applied to adjacent floors. Continuity of contact between the floor and wall is also acceptable.

[*Note*: Items 1 and 2 assume that joist spacing does not exceed 1.2 m and the house is not over two storeys. If over two storeys, supplementary strapping should be provided to joists in the longitudinal direction as shown in Fig. 4.5.1.]

SLENDERNESS RATIO

This is an acceptable standard for measuring dimensional stability. Actual wall height and thickness (see Chapters 3.1–3.3) can be varied for structural calculation purposes. This is known as the effective height or thickness and depends on the

Figure 4.5.1 Lateral restraint connections between wall and floor

Example applied to housing

Figure 4.5.2   Effective height (*H*) calculation

method (if any) of restraint. The effectiveness of connecting a wall to adjacent structure is shown in Fig. 4.5.2.

Calculations to determine effective wall thickness are also possible for different wall constructions. Single leaf/solid walls without piers have an effective thickness equal to the actual thickness. Attached piers will cause some variation and reference to BS 5628: Part 1: *Structural use of unreinforced masonry*, will provide appropriate

Figure 4.5.3 Effective length (*L*) calculation

coefficients. Cavity walls have an effective thickness equal to two-thirds of the sum of the actual thickness of both leaves, or the actual thickness of the thicker leaf (whichever is greater).

Effective length can also be important for structural calculations. This depends on the extent of support from abutments as shown in Fig. 4.5.3.

CALCULATION OF SLENDERNESS RATIO EXAMPLE

A cavity wall consists of 102.5 mm brick outer leaf and 150 mm concrete block inner leaf. The wall height from floor to ceiling is 3 m, with floor joists built into the inner leaf (enhanced support).

Effective thickness $= 0.67\ (102.5 + 150) = 169$ mm

Effective height $\quad = 0.75 \times 3{,}000 = 2{,}250$ mm

$$\text{Slenderness ratio} = \frac{\text{effective height}}{\text{effective thickness}}$$

$$= \frac{2{,}250}{169} = 13.30$$

Maximum acceptable slenderness ratios vary with applications in accordance with BS 5628: Part 1: *Structural use of unreinforced masonry*. The main objective is to evaluate and compare numerically different construction methods. The lower the slenderness ratio, the less prone a structure is to buckling under load due to insufficient rigidity.

PART 5

FUEL COMBUSTION

FIREPLACES, CHIMNEYS AND FLUES

The traditional method of providing heating in a domestic building is the open fire burning coal and/or wood but because of its low efficiency compared to modern heating appliances its use as the primary source of heating has declined. It must be noted that all combustible fuels, such as coal, smokeless fuels, oils, gas and wood, require some means of conveying the products of combustion away from the appliance or fireplace to the open air. They must discharge in a manner which does not create an environmental hazard or fire risk.

Terminology

Fireplace   This is the area in which combustion of the fuel takes place. It may be in the form of an open space with a fret or grate in which wood or coal is burnt, or a free standing appliance such as a slow combustion stove, room heater, oil-burning appliance, gas fire or gas boiler.

Chimney   The fabric surrounding the flue and providing it with the necessary strength and protection.

Flue   This is strictly a void through which the products of combustion pass. It is formed by lining the inside of a chimney with a suitable lining material to give protection to the chimney fabric from the products of combustion, as well as forming a flue of the correct shape and size to suit the type of fuel and appliance being used.

HEAT-PRODUCING APPLIANCES

Building Regulations J1 to J3 are concerned with heat-producing appliances and require that they are installed with an adequate air supply for efficient working, provided with a suitable means of discharging the products of combustion to the outside air and installed so as to reduce the risk of the building catching fire. Approved Document J gives practical guidance to satisfy these regulations and makes particular reference to solid-fuel and oil-burning appliances with rated outputs up to 50 kW and 45 kW respectively. The Document also refers to gas-burning appliances with rated input up to 70 kW. Input and output ratings can be used to calculate the efficiency of appliances, e.g. if a gas fire has input and output ratings of 12 kW and 9 kW respectively, the appliance efficiency is

$$\frac{9}{12} \times \frac{100}{1} = 75\%$$

When a fuel is burnt to provide heat it must have an adequate air supply to provide the necessary oxygen for combustion to take place. The air supply is drawn through the firebed in the case of solid fuels and injected in the case of oil-burning appliances. A secondary supply of air is also drawn over the flames in both cases. In rooms where an open fire is used the removal of air induces ventilation and this helps to combat condensation. The air used in combustion is drawn through the appliance by suction provided by the flue and this is affected by the temperature, pressure, volume, cross-sectional area and surface lining of the flue.

Openings for air entry to rooms containing appliances must satisfy the following:

1. Solid-fuel open appliance – have an area at least 50% of the appliance throat opening.
2. Other solid-fuel and oil-burning appliances (unless fitted with a balanced flue) – have an area at least 550 mm² per kW of rated output above 5 kW. If a flue draught stabilizer is fitted to a solid-fuel appliance, an extra 300 mm² is required for each kW of rated output.
3. Gas appliances with an open flue – have an area at least 500 mm² per kW of input rating over 7 kW.
4. Gas appliances with room-sealed-balanced flues do not require special provision, but gas fires and solid-fuel-effect fires should have ventilation as for solid-fuel appliances or in accordance with BS 5871: Parts 1 to 3.

Note: Further details of gas appliance flue installation are considered in the next chapter.

To improve the efficiency of a flue it is advantageous for the chimney to be situated on an internal wall since this will reduce the heat losses to the outside air, also the flue gas temperature will be maintained and this will, in turn, reduce the risk of condensation within the flue. The condensation could cause corrosion of the chimney fabric due to the sulphur compounds which are a by-product of the processes of combustion.

CHIMNEYS

A chimney should be built as vertical as is practicable to give maximum flue gas flow – if bends are required the angle should be 30° maximum with the vertical. If an appliance is connected to a chimney the flue pipe should be as short as possible and continuous with an access for cleaning included at the base of the main flue. The chimney should be terminated above the roof so that it complies with the Building Regulations and is unaffected by adverse pressures which occur generally on the windward side of a roof and may cause down draughts. The use of a chimney terminal or pot could be to transfer the rectangular flue section to a circular cross-section giving better flue flow properties or to provide a cover to protect the flue from the entry of rain or birds. A terminal should be fixed so that it does not impede in any way the flow of flue gases. The construction should be from solid non-combustible materials capable of withstanding temperatures of 1,100 °C, without impairing stability. Brickwork and blockwork must be lined with suitable materials as defined in Approved Document J, Sections 1.27 and 1.28. Unlined blockwork chimneys of refractory material or a combination of high-alumina cement and kiln burnt or pumice aggregates are acceptable.

BUILDING REGULATIONS

Combustion appliances and fuel storage systems are covered by Part J of the Regulations, with Approved Documents in 6 sections. The main requirements for solid-fuel are shown in Figs. 5.1.1–5.1.4.

Firebacks

The functions of a fireback are to contain the burning fuel, to prevent the heat of the fire from damaging the wall behind it and to radiate heat from the fire into the room. The space behind a fireback should be of a solid material though not a strong mix and not composed of loose rubble; the recommended mix is 1:2:4 lime, sand and broken brick, alternatively an insulating mix of 5 parts of vermiculite to 1 part of cement could be used. The temperature encountered by a fireback will be high, therefore the problem of expansion and subsequent contraction must be considered. The fireback should preferably be in two or more parts since the lower half will become hotter than the upper half. Multi-piece firebacks also have the advantage of being easier to fit. It is also good practice to line the rear of a fireback with corrugated paper or a similar material which will eventually smoulder away leaving a small expansion gap at the back of the fireback.

Throat

The size and shape of the throat above the fireplace opening is of the utmost importance. A fireplace with an unrestricted outlet to the fire would create unpleasant draughts by drawing an unnecessary amount of air through the flue and reduce the efficiency of the fire by allowing too much heat to escape into the flue. A throat restriction of 100 mm will give reasonable efficiency without making chimney sweeping impossible (see Fig. 5.1.5).

Figure 5.1.1   Chimney construction

Weathering or flaunching in cm. mt. (1:3)

Chimney terminal or pot

600 min. (Max. 4.5 times least lateral dimension)

Header oversailing course

Code 4 lead stepped flashing

Typical outlet detail

To comply with AD, J, Section 2.3

For maximum draw flues to be as straight as possible

238 dia.

150 dia.

295 or 372

Flue gathering

Angle between 45° & 60°

300 to 900

13

Angle 30° max to vertical

19

585

200 dia.

Rebated flue liner

Chimney pot

Recess 328 deep

580

Figure 5.1.2   Typical flue construction

Flue liners
to BS 1181

Back of recess to
minimum thickness
for full height of
recess

BS 1251
lintel

Jambs to comply with
AD, J, Section 2.20

Superimposed hearth

Non-combustible
DPC solidly bedded
in mortar

Constructional hearth
not less than 125 thick –
AD, J, Section 2.18

Cavity wall

100
min.

100
min.

200
min.

200
min.

Minimum dimensions
for constructional hearths
– AD, J, Section 2.18

Separating wall

200
min.

500

150   Recess   150

100
min.

100
min.

External wall and
non-combustible
external cladding

100
min.

Other than separating wall

Figure 5.1.3   Fireplaces and Approved Document J

Upper floor fireplaces – J2

Figure 5.1.4 Fireplaces and outlets – Approved Document J

Surrounds

This is the façade of a fireplace and its main function is an attractive appearance. It can be of precast concrete with an applied finish such as tiles or built *in-situ* from small brickettes or natural stones. If it is precast it is usually supplied in two pieces, the front and the hearth. The front is fixed by screws through lugs cast into the edges of the surround and placed against a 25 mm wide non-combustible cord or

Open fireplaces

Figure 5.1.5 Typical traditional domestic fireplaces

rope around the fireplace opening to allow for expansion and contraction. The hearth should be bedded evenly on the constructional hearth with at least 10 mm of 1:1:8 cement:lime:sand mix (see Fig. 5.1.5).

DEEP ASHPIT FIRE

This is a low front fire with a fret or grate being at or just below hearth level. A pit is constructed below the grate to house a large ashpan capable of holding several days ash. The air for combustion is introduced through the ashpit and grate by means of a 75 mm diameter duct at the end of which is an air control regulator. If the floor is of suspended construction the air duct would be terminated after passing through the fender wall, whereas with a solid floor the duct must pass under the floor and be terminated beyond the external wall, the end being suitably guarded against the entry of vermin.

BACK BOILER

This an open fire of conventional design with part of the fireback replaced by a small boiler with a boiler flue and a control damper. These are rarely installed now, but were the main source of hot water supply for domestic premises prior to oil or gas central heating becoming more widespread in the 1960s.

OPEN FIRE CONVECTORS

These are designed to increase the efficiency of an open fire by passing back into the room warm air as well as the radiated heat from the burning fuel. The open convector is a self-contained unit consisting of a cast-iron box containing the grate and fireback forming a convection chamber. The air in the chamber heats up and flows into the room by convection currents moving the air up and out of the opening at the top of the unit.

ROOM HEATERS

These are similar appliances to open fire convectors but they are designed to burn smokeless fuels and operate as a closed unit. In some models the strip glass front is, in fact, a door and can therefore be opened to operate as an open fire. Room heaters are either free standing, that is, fixed in front of the surround with a plate reducing the size of the fireplace opening to the required flue pipe size, or inset into the fireplace recess where the chimney flue aperture is reduced by a plate to receive the flue pipe of the appliance.

INDEPENDENT BOILERS

The function of these appliances is to heat water whether it is for a central heating system or merely to provide hot water for domestic use. It is generally fixed in the kitchen because of the convenience to the plumbing pipe runs and to utilize the background heat emitted. It must discharge into its own flue since it is a slow combustion appliance and the fumes emitted could, if joined to a common flue, cause a health hazard in other rooms of the building. Adequate access must be made for sweeping the flue and removing the fly ash that will accumulate with a smokeless fuel (see Fig. 5.1.5). Further details of boiler flues are considered in the next chapter.

BOILER FLUES

Boilers are appliances designed to heat water by energy exchange from burning oil, gas or solid fuels. Application of boilers to domestic systems of hot water circulation and distribution are provided in Chapter 10.1.

The term *boiler* is something of a misnomer, originating from Victorian times when appliances were used to raise steam for heating and industrial processes. The modern domestic boiler in an open-vented system is designed to heat water to about 80 °C. Thereafter it is uneconomic and unsafe. They are either floor standing with a conventional open flue pipe, possibly using an existing chimney or wall mounted with a room-sealed-balanced flue. Wall mounting permits discrete positioning between or within kitchen units or other unobtrusive location. Floor-standing boilers are now less common for domestic use, as the technology has advanced sufficiently to provide more compact units for wall mounting with simple horizontal balanced flue.

Floor-standing boilers with conventional open flue incorporate an air opening from the room containing the boiler to aid combustion. The area of ventilation opening is considered in the previous chapter. Flue installation must allow for positioning of the terminal (outlet), flue length and size. The simplest installation for a gas boiler is a straight vertical pipe as shown in Fig. 5.2.1, with non-combustible sleeves provided where it passes through a floor and other combustible parts of the structure. The advantage of a simple straight flue is less surface exposure than a flue offset to the outer wall as in Fig. 5.2.2. Here there is more opportunity for condensation in the flue, resulting from flue gases cooling to the dew point of water – approx. 60 °C.

Figure 5.2.1　Internal open flue

— Terminal

600 min.
above openable
window or vent

Sockets uppermost ⟶

Secondary flue

135° min. bend

Condensate pipe ⟶
to gulley if required

600 min.

Draught diverter

Primary flue

Air inlet ⟶

Boiler

Figure 5.2.2   Traditional open flue

FLUE MATERIALS

Flue materials are detailed in the Building Regulations, Approved Document J.
They can incorporate traditional brick chimneys, precast concrete flue blocks (gas
appliances only) or flue pipes of various materials. Chimneys are unnecessarily
large for domestic boiler purposes, but may be used with clay liners as defined in
the preceding chapter. In older premises where linings were omitted, the chimney
may receive a flexible stainless steel lining inserted as shown in Fig. 5.2.3. These
are not suitable for use with solid-fuel appliances as the products of combustion
can deteriorate the lining.

Precast concrete flue blocks to BS 1289 are manufactured from high-alumina
cement and dense aggregates, to resist chemical decomposition from acidity
caused by flue gases and condensation. They are jointed with high-alumina cement
mortar and incorporate a continuous void for the gases to escape. Dimensions
co-ordinate with the inner leaf of blockwork as shown in Fig. 5.2.4. The minimum

Figure 5.2.3   Flexible stainless steel flue lining

cross-sectional flue area is 16,500 mm² for conveyance of flue gases from gas fires, convectors and other small appliances.

Connecting flue pipes are produced in a variety of materials to suit application. For solid-fuel and oil-burning appliances (up to 50 kW and 45 kW output respectively) with flue gas temperatures likely to exceed 250°C; flue pipes may be of enamelled cast iron to BS 41, 3 mm minimum thickness mild steel to BS 1449, enamelled steel to BS 6999 or stainless steel as described in BS EN 10088-1. Gas appliances (up to 70 kW input) and oil-burning appliances (up to 45 kW output) with flue gas temperatures unlikely to exceed 250°C, may use the preceding materials or lighter weight connecting flue pipes of sheet metal to BS 715 or fibre cement to BS 7435.

Double wall steel variations of annular construction (pipe within a pipe) are preferred, as the space between each pipe can be filled with insulation against heat transfer to the adjacent structure or the open air. The insulation will also reduce condensation within the flue.

Flue terminals or outlets should promote extraction of flue gases, resist potential downdraughts and prevent the ingress of rain and snow. For solid-fuel and oil

External cavity wall

Figure 5.2.4 Precast concrete flue blocks

appliances, see the previous chapter (Fig. 5.1.4). For gas appliances flue terminals are shown in Fig. 5.2.5, with preferred location at or above the ridge of a pitched roof or at least 600 mm above the intersection with the roof pitch.

Balanced flues are specifically suited to wall-mounted boilers. The air for combustion is drawn directly into the boiler through the terminal in the external wall, therefore no special provision for room ventilation is necessary. The burnt gases discharge through the combined flue/intake unit as shown in Fig. 5.2.6 to maintain a constant balanced air pressure at the gas burner, regardless of outside wind conditions. Some flues incorporate a fan for more efficient control of combustion gases. This also permits a longer horizontal flue where it is impractical to mount a boiler directly on an external wall.

Balanced flue location is important for efficient operation of the appliance, safe discharge of flue gases and minimizing heat damage to adjacent structure and components. Table 5.2.1 lists acceptable locations for the flues and Fig. 5.2.7 illustrates these requirements.

Pipe terminals

Ridge terminal

Figure 5.2.5   Flue terminals

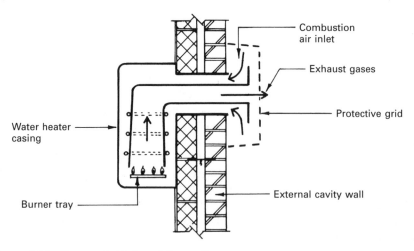

Figure 5.2.6   Room sealed balanced flue appliance

Figure 5.2.7   Balanced flue terminal positions – minimum dimensions (gas appliances)

Table 5.2.1   Minimum dimensions for location of balanced flue terminals

Position	Minimum distance (mm)	
	Natural draught	Fanned draught
Under an openable window or ventilator	300	300
Under rainwater goods or sanitation pipework	300	75
Under eaves	300	200
Under a balcony or a car port roof	600	200
From a window or door in the car port	1,200	1,200
Horizontally from vertical drain and soil pipes	300	150
Horizontally from internal or external corners	600	300
Above ground level or a projection, e.g. balcony	300	300
From an opposing surface	600	600
From an opposing terminal	600	1,200
Vertically from another terminal	1,500	1,500
Horizontally from another terminal	300	300

PART 6

ROOFS

ROOFS: TIMBER, FLAT AND PITCHED

TIMBER FLAT ROOFS

A flat roof is essentially a low-pitched roof and is defined in BS 6100: *Glossary of building and civil engineering terms*, as a pitch of 10° or less to the horizontal. Generally the angle of pitch is governed by the type of finish which is to be applied to the roof.

The functions of any roof are:

1. To keep out rain, wind, snow and dust.
2. To prevent excessive heat loss in winter.
3. To keep the interior of the building cool in summer.
4. Designed to accommodate all stresses encountered.
5. Designed to accept movement due to changes in temperature and moisture content.
6. To provide lateral restraint and stability to adjacent walls.
7. To resist penetration of fire and spread of flame from external sources.

The simplest form of roof construction to fulfil these functions is a timber flat roof covered with an impervious material to prevent rain penetration. This form of roof is suitable for spans up to 4,000 mm, spans over this are usually covered with a reinforced concrete slab or a patent form of decking.

The disadvantages of timber flat roofs are:

- They have limited capacity for insulation.
- They may contrast in style with other buildings in the vicinity and if an extension, the building to which it is attached.
- Unless properly designed and constructed, pools of water will collect on the surface causing local variations in temperature. This results in deterioration of the covering and, consequently, high maintenance costs.
- They have little or no space to accommodate services.

CONSTRUCTION

The construction of a timber flat roof follows the same methods as those employed for the construction of timber upper floors. Suitable joist sizes can be obtained by design or by reference to tables in publications recommended in the Building Regulations, Approved Document A. The spacing of roof joists is controlled by the width of decking material to be used and/or the width of ceiling board on the underside. Timber flat roofs are usually constructed to fall in one direction towards a gutter or outlet. This can be achieved by sloping the joists to the required fall but, as this would give a sloping soffit on the underside, it is usual to fix wedge shaped fillets called 'firrings' to the top of the joists to provide the fall (see Fig. 6.1.1). The materials used in timber flat roof construction are generally poor thermal insulators and therefore some form of non-structural material can be incorporated into the roof if it has to comply with Part L of the Building Regulations.

Decking materials

Timber   This can be in the form of softwood boarding, chipboard, oriented strand board or plywood. Plain edge sawn softwood boards or tongued and grooved boards are suitable for joists spaced at centres up to 450 mm. Exterior grade water- and boil-proof (WBP) plywood is available in sheet form which requires fixing on all four edges. This means noggins will be required between joists to provide the bearing for end fixings. Chipboard is also a suitable sheet material and is fixed in a similar manner to plywood. This material can be susceptible to moisture movement, therefore for roofing it should be specified 'moisture resistant to BS 5669: Part 2, type C3, C4 or C5'. Flat roofs must have the void between ceiling and decking ventilated to prevent condensation occurring. It is also advisable to use structural timbers which have been treated against fungal and insect attack. In certain areas treatment to prevent softwood infestation by the house longhorn beetle is a requirement under the Building Regulations. Details can be found in the Approved Document to support Regulation 7.

Compressed straw slabs   These are made from selected straw by a patent method of heat and pressure to a standard width of 1,200 mm × 2,400 mm length, the standard thickness being 58 mm which gives sufficient strength for the slabs to span 600 mm. All edges of the slabs must be supported and fixed. Ventilation is of the utmost importance and it is common practice to fix cross-bearers at right angles to and over the joists to give cross-ventilation. A bitumen scrim should be placed over the joints before the weathering membrane is applied.

Wood wool slabs   These are 610 mm wide slabs of various lengths which can span up to 1,200 mm. Thickness varies, for roof decking 51 mm is normally specified. The slabs are made of shredded wood fibres which have been chemically treated and are bound together with cement. The fixing and laying is similar to compressed straw slabs.

Insulating materials

There are many types of insulating materials available, usually in the form of boards or quilts. Insulation boards laid over the decking create a 'warm deck' roof, whereas quilted materials draped over the joists or placed between them, make a 'cold deck'.

Boards   These originated from lightly compressed vegetable fibres, bonded with natural glues or resins. Compressed straw and wood wool slabs have been particularly successful as decking materials with the benefit of in-built insulation.

For insulation only purposes, the most popular materials for roofing boards are high-density mineral wool with a tissue membrane bonded to one surface, expanded polystyrene or polyurethane slabs. A warm deck roof has insulation board placed over the decking, but below the water-proof membrane. Conversely, an inverted warm deck has the insulation board above the water-proof membrane. The inverted warm deck insulation board must be unaffected by water and be capable of receiving a surface treatment of stone granules or ceramic pavings. With this technique, the water-proof membrane is protected from the stresses caused by exposure to weather extremes. Examples of the various forms of construction are shown in Fig. 6.1.3.

Quilts These are made from mineral or glass wool which is loosely rolled with the option of a kraft paper facing. The paper facing is useful as the wool is in fine shreds giving rise to irritating scratches if handled. Quilts rely on the loose way in which the core is packed for their effectiveness and therefore the best results are obtained when they are laid between joists.

A variety of loose fills are also available for placing between the joists and over the ceiling to act as thermal insulators. Existing flat roofs can have their resistance to heat loss improved by applying thermal insulation laminated plasterboard to the ceiling.

Weather-proof finishes

Suitable materials are asphalt, lead, copper, zinc, aluminium and built-up roofing felt; only the latter will be considered at this stage.

Built-up roofing felt

Most roofing felts consist essentially of a base sheet of glass fibre or polyester reinforcement, impregnated with hot bitumen during manufacture. This is coated on both sides with a stabilized weatherproof bitumen compound. The outer coating is dusted with sand while still hot and tacky. The under layer may receive a thin layer of polythene to prevent the sheet sticking to itself when rolled. After cooling the felt is cut to form rolls 1 m wide and 10 or 20 m long before being wrapped for dispatch.
BS 747: *Specification for roofing felts*, contains two types:

■ Type 3: Glass fibre reinforced.
■ Type 5: Polyester reinforced.

Both types have subdivisions relating to the surface finish. Other bituminous sheet materials are available which incorporate polymer modifications to the bitumen. These are styrene–butadiene–styrene (SBS) and atactic–polypropylene (APP). SBS improves the physical properties and APP has a higher melting point, most suited to torch-on applications. Rag fibre reinforced bituminous felts are still available, but no longer recommended for roofing habitable accommodation.
For flat roofs three layers of felt should be used, the first being laid at right angles to the fall commencing at the eaves. If the decking is timber the first layer

is secured with large flat-head felt nails and the subsequent layers are bonded to it with a hot bitumen compound by a roll and pour method. This involves pouring molten bitumen on the decking or underlayer and unrolling the sheet over it. Torch-on is for specially made sheets which are heated to the underside to produce a wave of molten bitumen while the sheet is unrolled. If the decking is of a material other than timber, all three layers are bonded with hot bitumen compound. It is usually recommended that a vented first layer is used in case moisture is trapped during construction; this recommendation does not normally apply to roofs with a timber deck since timber has the ability to 'breathe'. The minimum fall recommended for built-up roofing felt is 17 mm in 1,000 mm or 1°.

In general the Building Regulations require a flat roof with a weatherproofing which has a surface finish of asbestos-based bituminous felt or is covered with a layer of stone chippings. The chippings protect the underlying felt, provide additional fire resistance and give increased solar reflection. A typical application would be 12.5 mm stone chippings at approximately 50 kilograms to each 2.5 square metres of roof area. Chippings of limestone, granite and light-coloured gravel would be suitable.

Vapour control layer/membrane

The problem of condensation should always be considered when constructing a flat roof. The insulation below the built-up roofing felt will not prevent condensation occurring and since it is a permeable material water vapour will pass upwards through it and condense on the underside of the roofing felt. The drops of moisture so formed will soak into the insulating material, lowering its insulation value and possibly causing staining on the underside. To prevent this occurring a vapour control layer should be placed on the underside of the insulating material. Purpose-made plasterboard with a metallized polyester backing can be specified in this situation, to combine water vapour resistance with the ceiling lining. The inverted warm deck roof will not require a separate vapour control layer, as the waterproof membrane will provide this function. For typical timber flat roof details, see Figs 6.1.1 and 6.1.2.

Firring pieces

Figure 6.1.1   Timber flat roof slope details

Stone chippings

3 layer built-up roofing felt

Decking

Firring

Joist

Fascia

Eaves ventilation

Cold deck: eaves detail

Insulation quilt between joists

Vapour control layer

Plasterboard

Cement particle board cavity seal

Insulating roof board

Code 4 lead apron

Stepped DPC cavity tray

VCL

Joist

Angle fillet

Warm deck: abutment detail

Inverted insulating roof board

Ceramic paving or pebble ballast

Waterproof membrane

Inverted warm deck

Figure 6.1.2   Timber flat roofs

TIMBER PITCHED ROOFS

The term *pitched roof* includes any roof whose angle of slope to the horizontal lies between 10° and 70°; below this range it would be called a flat roof and above 70° it would be classified as a wall.

The pitch is generally determined by the covering which is to be placed over the timber carcass, whereas the basic form is governed by the load and span. The terminology used in timber roof work and the basic members for various spans are shown in Figs 6.1.3 and 6.1.4.

The cost of constructing a pitched roof is greater than a flat roof, but the pitch will create a useful void for locating cold water storage cisterns and for general storage. The timber used in roof work is structural softwood, the members being joined together with nails. The sloping components or rafters are used to transfer the covering, wind, rain and snow loads to the loadbearing walls on which they rest. The rafters are sometimes assisted in this function by struts and purlins in what is called a purlin or double roof (see Fig. 6.1.5). As with other forms of roofs the spacing of the rafters and consequently the ceiling joists is determined by the module size of the ceiling boards which are to be fixed on the underside of the joists.

Roof members

Ridge   This is the spine of a roof and is essentially a pitching plate for the rafters which are nailed to each other through the ridge board. The depth of ridge board is governed by the pitch of the roof, the steeper the pitch the deeper will be the vertical or plumb cuts on the rafters abutting the ridge.

Common rafters   The main loadbearing members of a roof; they span between a wall plate at eaves level and the ridge. Rafters have a tendency to thrust out the walls on which they rest and this must be resisted by the walls and the ceiling joists. Rafters are notched over and nailed to a wall plate situated on top of a loadbearing wall; the depth of the notch should not exceed one-third the depth of the rafter.

Jack rafters   These fulfil the same function as common rafters but span from ridge to valley rafter or from hip rafter to wall plate.

Hip rafters   Similar to a ridge but forming the spine of an external angle and similar to a rafter spanning from ridge to wall plate.

Valley rafters   As hip rafter but forming an internal angle.

Wall plates   These provide the bearing and fixing medium for the various roof members and distribute the loads evenly over the supporting walls; they are bedded in cement mortar on top of the load-bearing walls.

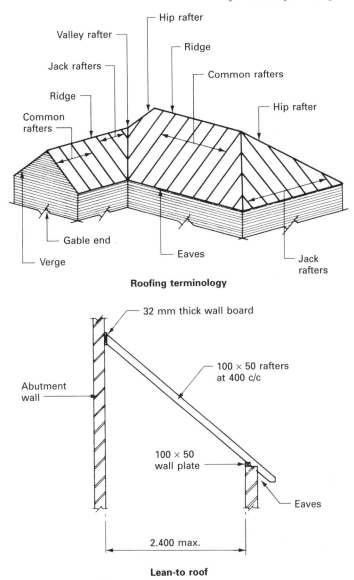

Figure 6.1.3 Roofing terminology and lean-to roof

Dragon ties A tie placed across the corners and over the wall plates to help provide resistance to the thrust of a hip rafter.

Ceiling joists These fulfil the dual function of acting as ties to the feet of pairs of rafters and providing support for the ceiling boards on the underside and any cisterns housed within the roof void.

Figure 6.1.4   Pitched roofs for small spans

Purlins   These act as beams reducing the span of the rafters enabling an economic section to be used. If the roof has a gable end they can be supported on a corbel or built in but in a hipped roof they are mitred at the corners and act as a ring beam.

Struts   These are compression members which transfer the load of a purlin to a suitable loadbearing support within the span of the roof.

Collars   These are extra ties to give additional strength and are placed at purlin level.

Binders   These are beams used to give support to ceiling joists and counteract excessive deflections and are used if the span of the ceiling joist exceeds 2,400 mm.

Hangers   Vertical members used to give support to the binders and allow an economic section to be used; they are included in the design if the span of the binder exceeds 3,600 mm.

[*Note*: The arrangement of struts, collars and hangers only occurs on every fourth or fifth pair of rafters.]

EAVES

The eaves of a roof is the lowest edge which overhangs the wall thus giving the wall a degree of protection; it also provides the fixing medium for the rainwater gutter. The amount of projection from the wall of the eaves is a matter of choice but is generally in the region of 300–450 mm.

There are two basic types of eaves finish: open eaves and closed eaves. The former is less expensive, with rafters left exposed on the underside and treated with preservative. With both, the space between the rafters and the roof covering receives insulation over the top of the wall. A 50 mm air space must remain above the insulation for free air circulation through the roof in order to prevent condensation occurring. This is easily achieved with a proprietory eaves ventilator secured between the rafters. A continuous triangular tilting fillet is fixed over the backs of the rafters to provide support for the bottom course of slates or tiles. A closed eaves is where the feet of the rafters are boxed in using a vertical fascia board, with the space between the fascia and the wall containing a ventilated soffit board. In a cheaper variant, the rafters are cut marginally beyond the wall face to leave space for ventilation and only a fascia board fixed to the rafter ends. This is called a *flush* eaves. Figures 6.1.5 and 6.1.6 indicate various rafter finishes, with provision for adequate through ventilation.

TIMBER ROOF TRUSSES

These can be used on the larger spans in domestic work to give an area below the ceiling level free from loadbearing walls. Trusses are structurally designed frames based on the principles of triangulation and serve to carry the purlins; they are

100 × 50 rafters at 400 c/c

32 thick ridge board

100 × 50 rafters at 400 c/c

Alternative purlin arrangement

Struts, collars and hangers to every 4th pair of rafters

Ventilation spacer

Tilting fillet

25

100 × 50 collar

Open eaves detail

100 × 50 rafters at 400 c/c

150 × 63 purlin

100 × 50 struts

100 × 50 hangers

100 × 50 binder

100 × 50 wall plate bedded in cm. mt. (1:3)

Loadbearing partition

Insulated external cavity wall

Fascia board

100 × 50 wall plate

Cement particle board cavity closer

Ventilated soffit to closed eaves, fixed to cradling

Note: Mineral wool insulation (or equivalent) placed between and over ceiling joists or between rafters. Thickness in accordance with Building Regulations, Approved Document L

Figure 6.1.5   Typical double or purlin roof details for spans up to 7,200 mm

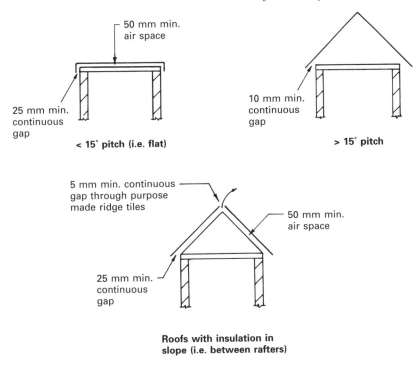

50 mm min.
air space

25 mm min.
continuous
gap

< 15° pitch (i.e. flat)

10 mm min.
continuous
gap

> 15° pitch

5 mm min. continuous
gap through purpose
made ridge tiles

50 mm min.
air space

25 mm min.
continuous
gap

**Roofs with insulation in
slope (i.e. between rafters)**

Note: Gaps need not be continuous, if staggered voids provide the equivalent area

Figure 6.1.6 Roof ventilation requirements

spaced at 1,800 mm centres with the space between being infilled with common rafters. It is essential that the members of a roof truss are rigidly connected together since light sections are generally used. To make a suitable rigid joint bolts and timber connectors are used; these are square- or circular-toothed plates, the teeth being pointed up and down which, when clamped between two members, bite into the surface forming a strong connection and spread the stresses over a greater surface area. A typical roof truss detail is shown in Fig. 6.1.7.

TRUSSED RAFTERS

This is another approach to the formation of a domestic timber roof giving a clear span; as with roof trusses it is based upon a triangulated frame but in this case the members are butt jointed and secured with truss plates. All members in a trussed rafter are machined on all faces so that they are of identical thickness ensuring a strong connection on both faces. The trussed rafters are placed at 600 mm centres and tied together over their backs with 38 × 25 tiling battens; no purlin or ridge is required. Stability is achieved from the tile battens and 100 × 25 diagonal wind braces from the bottom corner to a top corner. Also, two 100 × 25 longitudinal ties should run horizontally along the ceiling ties.

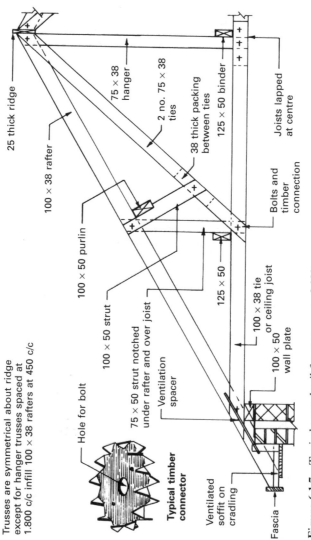

Trusses are symmetrical about ridge
except for hanger trusses spaced at
1.800 c/c infill 100 × 38 rafters at 450 c/c

25 thick ridge

100 × 38 rafter

75 × 38
hanger

2 no. 75 × 38
ties

38 thick packing
between ties

125 × 50 binder

Joists lapped
at centre

100 × 50 purlin

100 × 50 strut

Bolts and
timber
connection

Hole for bolt

75 × 50 strut notched
under rafter and over joist

Ventilation
spacer

125 × 50

100 × 38 tie
or ceiling joist

100 × 50
wall plate

**Typical timber
connector**

Ventilated
soffit on
cradling

Fascia

Figure 6.1.7 Typical truss detail for spans up to 8,000 mm

Truss manufacturers will design and prefabricate to the client's specification. This should include span, loading (type of tile), degree of exposure, pitch, spacing (if not 600 mm) and details of any special loadings such as water cistern location.

Truss or nail plates are generally of one or two forms:

1. Those in which holes are punched to take nails and are suitable for site assembly using a nailing gun.
2. Those in which teeth are punched and bent from the plate and are used in factory assembly using heavy presses.

In all cases truss plates are fixed to both faces of the butt joint.

Trussed rafters are also produced using gusset plates of plywood at the butt joints instead of truss plates; typical details of both forms are shown in Fig. 6.1.8.

LATERAL BRACING

Gable ladders and restraint straps secure pitched roofs to adjacent walls which, in turn, add stability to the wall. Figure 6.1.9 shows the application with straps at 2 m spacing. Straps may need to be more frequent, depending on building height and degree of exposure.

BUILDING REGULATIONS

Regulation 7 This requires that any building work shall be carried out with proper materials and in a workmanlike manner. The Approved Document supporting this regulation defines acceptable levels of performance for materials which include products, fittings, items of equipment and back-filling for excavations. The aids for establishing fitness of materials suggested in the Approved Document are:

1. Past experience – such as a building in use.
2. Agrément certificates or equivalent European Technical Approvals.
3. British Standards and European Community national standards.
4. Independent certification schemes to the National Accreditation Council for Certification Bodies (NACCB) requirements.
5. Quality assurance schemes complying with BS 5750 and BS EN ISO 9000–9004.
6. Test and calculations for materials only using the National Measurement Accreditation Service (NAMAS) scheme for testing laboratories.

The Approved Document supporting Regulation 7 lists certain geographical areas where the softwood timber used for roof construction should be adequately treated with a suitable preservative to prevent infestation by the house longhorn beetle. This is a wise precaution whether it is recommended or not.

Regulation B4 (Section 14) This requires the roof to offer adequate resistance to the spread of fire over the roof. Table 17 in Approved Document B gives

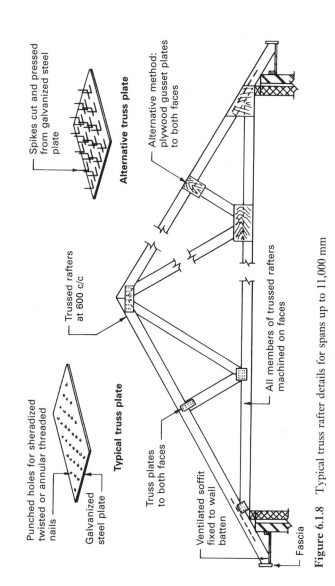

Spikes cut and pressed from galvanized steel plate

Alternative truss plate

Alternative method: plywood gusset plates to both faces

Trussed rafters at 600 c/c

All members of trussed rafters machined on faces

Punched holes for sheradized twisted or annular threaded nails

Galvanized steel plate

Typical truss plate

Truss plates to both faces

Ventilated soffit fixed to wall batten

Fascia

Figure 6.1.8 Typical truss rafter details for spans up to 11,000 mm

Figure 6.1.9   Lateral restraint to gable walls

limitations on roof coverings for dwelling houses by designations and distance from the boundary, the designations being defined in Table A5.

Regulation C2   This requires that the roof of a building shall adequately resist the passage of moisture to the inside of the building.

Regulation L1   This states that reasonable provision shall be made for the conservation of fuel and power in buildings. To satisfy this requirement Approved Document L gives maximum permitted U values for roofs of between 0.16 and 0.25 W/m^2K depending on application. This standard will not be achieved with traditional roof construction alone, therefore enhancement with suitable insulation material will be required. (See Chapter 8.2 for more detail.)

The selection of suitable structural timber members for pitched roofs of various loadings and spans can be determined by reference to tables from publications in Approved Document A of the Building Regulations.

ROOF TILING AND SLATING

A roof can be defined as the upper covering of a building and in this position it is fully exposed to the rain, snow, wind, sun and general atmosphere; therefore the covering to any roof structure must have good durability to meet these conditions. Other factors to be taken into account are weight, maintenance and cost.

Roofs are subjected to wind pressures, both positive and negative, the latter causing uplift and suction which can be overcome by firmly anchoring lightweight coverings to the structure or by relying upon the deadweight of the covering material. Domestic roofs are not usually designed for general access and therefore the chosen covering material should be either very durable or easily replaceable. The total dead load of the covering will affect the type of support structure required and ultimately the total load on the foundations, so therefore careful consideration must be given to the medium selected for the roof covering.

In domestic work the roof covering usually takes the form of tiling or slating since these materials economically fulfil the above requirements and have withstood the test of time.

TILING

Tiles are manufactured from clay and concrete to a wide range of designs and colours suitable for pitches from 15° to 45° and work upon the principle of either double or single lap. The vital factor for the efficient performance of any tile or slate is the pitch and it should be noted that the pitch of a tile is always less than the pitch of the rafters owing to the overlapping technique.

Tiles are laid in overlapping courses and rely upon the water being shed off the surface of one tile onto the exposed surface of the tile in the next course. The problem of water entering by capillary action between the tiles is overcome by the camber of the tile, the method of laying or by overlapping side joints. In all methods of tiling a wide range of fittings are produced to enable all roof shapes to be adequately protected.

PLAIN TILING

This is a common method in this country and works on the double lap principle. The tiles can be of hand-made or machine-pressed clay, the process of manufacture being similar to that of brickmaking. The hand-made tile is used mainly where a rustic or distinctive roof character is required since they have a wide variation in colour, texture and shape. They should not be laid on a roof of less than 45° pitch since they tend to absorb water and if allowed to become saturated they may freeze, expand and spall or fracture in cold weather. Machine-pressed tiles are harder, denser and more uniform in shape than hand-made varieties and can be laid to a minimum pitch of 35°.

A suitable substitute for plain clay tiles is the concrete plain tile. These are produced in a range of colours to the same size specifications as the clay tiles and with the same range of fittings. The main advantage of concrete tiles is their lower cost and the main disadvantage is the extra weight. Plain clay tile details are specified in BS 402 and BS EN 538. Concrete tiles have separate designation in BS EN 490 and 491.

Plain tiling in common with other forms of tiling provides an effective barrier to rain and snow penetration, but wind is able to penetrate into the building through the gaps between the tiling units, therefore a barrier in the form of boarding or sheeting is placed over the roof carcass before the battens on which the tiles are to be hung are fixed.

The rule for plain tiling is that there must always be at least two thicknesses of tiles covering any part of the roof and bonded so that no 'vertical' joint is immediately over a 'vertical' joint in the course below. To enable this rule to be maintained shorter length tiles are required at the eaves and the ridge; each alternate course is commenced with a wider tile of one-and-a-half tile widths. The apex or ridge is capped with a special tile bedded in cement mortar over the general tile surface. The hips can be covered with a ridge tile in which case the plain tiling is laid underneath and mitred on top of the hip or alternatively a special bonnet tile can be used where the plain tiles bond with the edges of the bonnet tiles. Valleys can be formed by using special tiles, mitred plain tiles or forming an open gutter with a durable material such as lead. The verge at the gable end can be formed by bedding plain tiles face down on the gable wall as an undercloak and bedding the plain tiles in cement mortar on the upper surface of the undercloak. The verge tiling should overhang its support by at least 50 mm. Abutments are made watertight by dressing a flashing over the upper surface of tiling between which is sandwiched a soaker. The soaker in effect forms a gutter. An alternative method is to form a cement fillet on top of the tiled surface but this method sometimes fails by the cement shrinking away from the surface of the wall.

The support or fixing battens are of softwood extended over and fixed to at least three rafters, the spacing or gauge being determined by the lap given to the tiles thus:

$$\text{Gauge} = \frac{\text{length of tile} - \text{lap}}{2}$$

$$\text{Gauge} = \frac{265 - 65}{2} = 100 \text{ mm}.$$

Plain tiles are fixed with two galvanized nails to each tile in every fourth or fifth course. Details of plain tiles, fittings and methods of laying are shown in Figs 6.2.1 to 6.2.5.

SINGLE-LAP TILING

Single-lap tiles are laid with overlapping side joints to a minimum pitch of 35° and are not bonded like the butt-jointed single-lap plain tiles; this gives an overall reduction in weight since less tiles are used. A common form of single-lap tile is the pantile which has opposite corners mitred to overcome the problem of four tile thicknesses at the corners (see Fig. 6.2.6). The pantile is a larger unit than the plain tile and is best employed on large roofs with gabled ends since the formation of hips and valleys is difficult and expensive. Other forms of single-lap tiling are Roman tiling, Spanish tiling and interlocking tiling. The latter types are produced in both concrete and clay and have one or two grooves in the overlapping edge to give greater resistance to wind penetration and can generally be laid as low as 15° pitch (see Fig. 6.2.6).

SLATING

Slate is a naturally dense material which can be split into thin sheets and used to provide a suitable covering to a pitched roof. Slates are laid to the same basic principles as double-lap roofing tiles except that every slate should be twice nailed. Slates come mainly from Wales, Cornwall and the Lake District and are cut to a wide variety of sizes – the Westmorland slates are harder to cut and are usually supplied in random sizes. Slates can be laid to a minimum pitch of 25° and are fixed by head nailing or centre nailing. Centre nailing is used to overcome the problem of vibration caused by the wind and tending to snap the slate at the fixing if nailed at the head; it is used mainly on the long slates and pitches below 35°.

The gauge of the battens is calculated thus:

$$\text{Head nailed gauge} = \frac{\text{length of slate} - (\text{lap} + 25 \text{ mm})}{2}$$

$$= \frac{400 - (75 + 25)}{2} = 150 \text{ mm}$$

$$\text{Centre nailed gauge} = \frac{\text{length of slate} - \text{lap}}{2}$$

$$= \frac{400 - 76}{2} = 162 \text{ mm}.$$

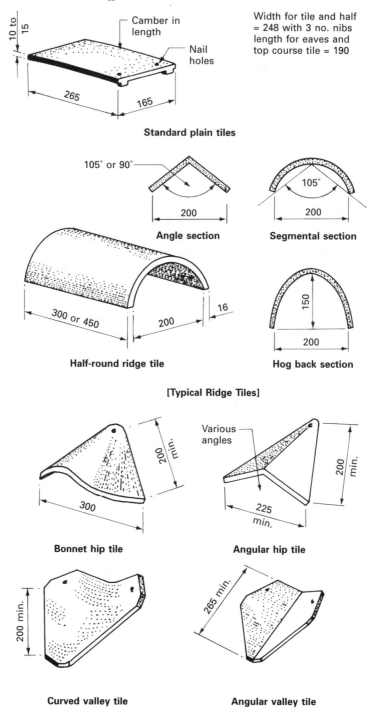

Standard plain tiles

Width for tile and half
= 248 with 3 no. nibs
length for eaves and
top course tile = 190

Angle section

Segmental section

Half-round ridge tile

Hog back section

[Typical Ridge Tiles]

Bonnet hip tile

Angular hip tile

Curved valley tile

Angular valley tile

Figure 6.2.1   Standard plain tiles and fittings

Alternative fixing for under-ridge tile – hung on 38 × 32 batten

Half-round ridge-capping bedded and jointed in cm. mt. (1:3)

Top course tile nailed to ridge

Ridge

Plain tiles laid face down as undercloak

Tile and half to complete bond

1:3 cm. mt. bed

Boxed fascia

Verge detail

Insulation

Gauge

Margin

Under-eaves course

Lap

Under-tiling felt

Insulated cavity wall

Cavity closer

Plain tiles on 38 × 19 battens, each tile in every 4th course nailed

Eaves course tile

Ventilated soffit

Figure 6.2.2 Plain tiling details

Typical abutment detail (see also Fig. 5.1.1)

Ridge termination at gable end

Ridge junction with hipped end

Figure 6.2.3   Abutment and ridge details

Bonnet hip tiles nailed
and bedded in cm. mt.
(1:3) to hip rafter

Tile slip

Plain tiles bonded with
bonnet hip tiles

Under-eaves tile course

Half-round hip capping
bedded in cm. mt. (1:3)

Galvanized hip
iron screwed to
top of hip rafter

Plain tiles mitred over hip
rafter under-hip capping

Figure 6.2.4   Hip treatments

Figure 6.2.5 Valley treatments

Side lap 32

Under-tiling felt

50 × 32 battens
at 268 c/c

End lap 75

Clay pantiles

Ends filled with
cm. mt. (1:3)

Under-eaves course
of plain tiles

40
Dishing

248

Corner
mitred

343

Roll

Pantile

229

381

Typical interlocking tile

Half-round ridge capping

3-way intersection

Half-round
hip tiles

Head
lap 81

Interlocking
tiles on 38 ×
32 battens of
300 c/c

Valley trough tiles

Figure 6.2.6   Examples of single lap tiling

Typical ridge detail

Head-nailed slating using 500 × 250 slates

Centre-nailed slating using 600 × 300 slates

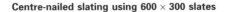

Figure 6.2.7   Typical slating details

Roofing slates are covered by BS 680 which gives details of standard sizes, thicknesses and quality. Typical details of slating are shown in Fig. 6.2.7.

UNDERTILING FELTS

The object of undertiling felt is to keep out dust and the wind. It also provides a secondary waterproof layer for limited protection against rain if tiles are damaged. It consists basically of a bituminous impregnated matted fibre sheet which can be reinforced with a layer of jute hessian embedded in the coating on one side to overcome the tendency of felts to tear readily. Undertiling felts are supplied in rolls 1 m wide and 10 or 20 m long depending upon type. They should be laid over the rafters and parallel to the eaves with 150 mm laps and temporarily fixed with large head felt nails until finally secured by the battens. Undertiling felts are also known as sarking felts and should conform to the requirements of BS 747. They should also be permeable to water vapour, to relieve any possibility of condensation in the roof space.

COUNTER BATTENS

If a roof is boarded before the roofing felt is applied the tiling battens will provide ledges on which dirt and damp can collect. To overcome this problem counter battens are fixed to the boarding over the rafter positions to form a cavity between the tiling battens and the boarding. Boarding a roof over the rafters is a method of providing a wind barrier and adding to the thermal insulation properties but is seldom used in new work because of the high cost.

ASPHALT FLAT ROOFS

Flat roofs are often considered to be a simple form of construction, but unless correctly designed and constructed they can be an endless source of trouble. Flat roofs can have the advantage of providing an extra area to a building for the purposes of recreation and an additional viewpoint. Mastic asphalt provides an ideal covering material especially where foot traffic is required.

Mastic asphalt consists of an aggregate with a bituminous binder which is cast into blocks ready for reheating on site (see Chapter 7.8). The blocks are heated in cauldrons or cookers to a temperature of over 200 °C and are then transported in a liquid state in buckets for application to the roof deck by hand spreading. Once the melted asphalt has been removed from the source of heat it will cool and solidify rapidly, therefore the distance between the cauldron and the point of application should be kept to a minimum. Mastic asphalt can be applied to most types of rigid substructure and proprietary structural deckings. To prevent slight movements occurring between the substructure and the finish an isolating membrane should be laid over the base before the mastic asphalt is applied.

Roof work will often entail laying mastic asphalt to horizontal, sloping and vertical surfaces and these are defined as follows:

1. Horizontal surfaces – up to 10° pitch.
2. Sloping surfaces – between 10° and 45° pitch.
3. Vertical surfaces – over 45° pitch.

The thickness and number of coats required will depend on two factors:

1. Surface type.
2. Substructure or base.

Horizontal surfaces with any form of rigid base should have a two–coat application of mastic asphalt laid breaking the joint and built up to a minimum total thickness of 20 mm. An isolating membrane of black sheathing felt complying with BS 747 Type 4A(i) should be used as an underlay laid dry over insulation boards with 50 mm lapped joints.

Vertical and sloping surfaces other than those with a timber base require a three-coat application built up to a 20 mm total thickness without an isolating membrane.

Timber substructures with vertical and sloping surfaces should have a three-coat 20 mm mastic asphalt finish applied to expanded metal lathing complying with BS 1369 fixed at 150 mm centres over an isolating membrane.

If the mastic asphalt surface is intended for foot traffic the total thickness of the two-coat application should be increased to a minimum of 25 mm.

Roofs with a mastic asphalt finish can be laid to falls so that the run off of water is rapid and efficient. Puddles of water left on the roof surface will create small areas of different surface temperatures which could give rise to local differential thermal movements and cause cracking of the protective covering. Depressions will also provide points at which dust and rubbish can collect; it is therefore essential that suitable falls are provided to direct the water to an eaves gutter or specific outlet points. The falls should be formed on the base or supporting structure to a gradient of not less than 1 in 80 (see typical details on Figs 6.3.1 and 6.3.2).

Alternatively the roof can be designed to act as a reservoir by a technique sometimes called *ponding*. The principle is to retain on the roof surface a 'pond' of water some 150 mm deep by having the surface completely flat, high skirtings and outlets positioned 150 mm above the roof level. The main advantage is that differential temperatures are reduced to a minimum; the disadvantages are the need for a stronger supporting structure to carry the increased load, a three-coat 30 mm thick covering of mastic asphalt and the need to flood the roof in dry hot weather to prevent the pond completely evaporating away.

Resistance to heat loss can be achieved by incorporating within the structure a dry insulation such as cork slabs, wood fibre boards, glass or mineral fibre boards, rigid polyisocyanurate (Pi) foam board or rigid lightweight expanded polystyrene (EPS) board. This may be deployed to create warm or cold deck roof insulating principles as described in the Chapter 6.1. If fixed over the roof structure it will reduce the temperature variations within the roof to a minimum and hence the risk of unacceptable thermal movements. A lightweight concrete screed will also prove a useful insulative supplement and a means of providing the roof slope. Suitable aggregates are furnace clinker, foamed blast furnace slag, pumice, expanded clay, sintered pulverized fuel ash, exfoliated vermiculite and expanded perlite. The thickness should not be less than 40 mm and it may be necessary to apply to the screed a 1:4 cement:sand topping to provide the necessary surface finish. Mix ratios of 1:8–10 are generally recommended for screeds of approximately 1,100 kg/m^3 density using foamed slag, sintered pulverized fuel ash and expanded clay, whereas a 1:5 mix is recommended for exfoliated vermiculite and expanded perlite aggregates giving a low density of under 640 kg/m^3. When mixing screeds containing porous aggregates high water/cement ratios are required to give workable mixes and therefore the screeds take a long time to dry out.

Moisture in the form of a vapour will tend to rise within the building and condense on the underside of the covering or within the thickness of the insulating material. If an insulating material becomes damp its efficiency will decrease and if composed of organic material it can decompose. To overcome this problem a vapour control layer of a suitable impermeable material such as polythene sheet or aluminium foil should be placed beneath the insulating layer. Care must be taken to

Figure 6.3.1 Timber flat roof with mastic asphalt covering

Labels on the figure:

- 12 mm ventilation gap
- Code 4 lead flashing
- 150 × 100 timber fillet
- 3-coat 20 mm thick asphalt
- 2-coat angle fillet
- Expanded metal lathing
- Vapour control layer
- Insulation between joists
- Joist hanger
- 9.5 mm plasterboard
- Firring
- 13 mm layer of light coloured stone chippings
- 2-coat 20 mm thick asphalt
- Isolating membrane
- Boarded deck
- 50 mm min. air space
- Metal trim
- Fascia
- Soffit
- 25 mm continuous ventilation gap

Figure 6.3.2 Reinforced concrete flat roof with mastic asphalt covering

Mineral wool insulation board

Metal trim

Timber fillet

13 mm layer of light-coloured stone chippings

Isolating membrane

2-coat 20 mm thick asphalt

Cavity insulation

Skirting dressed into 25 × 25 chase

2-coat asphalt skirting keyed to brickwork

2-coat angle fillet

150 min.

Lightweight insulating screed laid to falls

Vapour control layer

RC roof slab

see that all laps are complete and sealed and that the edges of the insulating material are similarly protected against the infiltration of moisture vapour. This is achieved by turning the vapour control layer around the edges of insulation board and bonding about 300 mm (min. 150 mm) to the top face of board.

The surface of a flat roof being fully exposed will gain heat by solar radiation and if insulated will be raised to a temperature in excess of the air or ambient temperature, since the transfer of heat to the inside of the building has been reduced by the insulating layer. The usual method employed to reduce the amount of solar heat gain of the covering is to cover the upper surface of the roof with light coloured chippings to act as a reflective finish. Suitable aggregates are white spar, calcined flint, white limestone or any light coloured granite of 12 mm size and embedded in a bitumen compound.

CHAPTER 6.4

LEAD-COVERED FLAT ROOFS

Lead as a building material has been used extensively for over 5,000 years and is obtained mainly from the mineral galena of which Australia, Canada, Mexico and the USA are the main producers. The raw material is mined, refined to a high degree of purity and then cast into bars or pigs which can be used to produce lead sheet, pipe and extruded products.

Lead is a durable and dense material $(11,340 \text{ kg/m}^3)$ of low strength but is very malleable and can be worked cold into complicated shapes without fracture. In common with other non-ferrous metals lead oxidizes on exposure to the atmosphere and forms a thin protective film or coating over its surface. When in contact with other metals there is seldom any corrosion by electrolysis and therefore fixing is usually carried out by using durable copper nails.

For application to flat roofs, milled lead sheet should comply with the recommendations of BS 1178. The sheet is supplied in rolls of widths varying between 150 mm and 2,400 mm, in 3 m and 6 m standard lengths. For easy identification, lead sheet carries a colour guide for each code number:

BS Code No.	Thickness (mm)	Weight (kg/m²)	Colour
3	1.32	14.97	green
4	1.80	20.41	blue
5	2.24	25.40	red
6	2.65	30.10	black
7	3.15	35.72	white
8	3.55	40.26	orange

The code number is derived from the former imperial notation of $5\text{lb/ft}^2 = \text{No. } 5$ lead.

The thickness or code number of lead sheet for any particular situation will depend upon the protection required against mechanical damage and the shape required. The following thickness can therefore be considered as a general guide for flat roofs:

Bossed end

Bossed end

Lead rolls at 600 to 800 c/c

2.250 to 2.750

Fall

Fascia

Lead clips at 750 c/c

Lead flashing turned into brick joint

150 mm high skirting

Firring

Vapour-check lined plasterboard

Ventilated soffit

Batten forming drip

Joist

Felt underlay

Decking of plywood chipboard or t & g boards laid diagonally

Figure 6.4.1 Typical layout of lead flat roof

1. Small areas without foot traffic, No. 4 or 5.
2. Small areas with foot traffic, No. 5, 6 or 7.
3. Large areas with or without foot traffic, No. 5, 6 or 7.
4. Flashings, No. 4 or 5.
5. Aprons, No. 4 or 5.

Milled lead sheet may be used as a covering over timber or similar deckings and over smooth screeded surfaces. In all cases an underlay of felt or stout building paper should be used to reduce frictional resistances, decrease surface irregularities and in the case of a screeded surface isolate the lead from any free lime present which may cause corrosion. Provision must also be made for the expansion and contraction of the metal covering. This can be achieved by limiting the area and/or length of the sheets being used. The following table indicates the maximum recommended area and length for any one piece of lead:

BS Code No.	Max. length between drips (m)	Max. area (m²)
3	Only suitable for soakers	—
4	1.50	1.13
5	2.00	1.60
6	2.25	1.91
7	2.50	2.25
8	3.00	3.00

Joints which can accommodate the anticipated thermal movements are in the form of rolls running parallel to the fall and drips at right angles to the fall positioned so that they can be cut economically from a standard sheet; for layout and construction details see Figs 6.4.1 and 6.4.2.

[*Note*: Current legislation will require these details to incorporate cavity insulation in the walls with an insulative concrete block inner leaf. Insulation must also be included within the roof construction as shown in Chapter 6.1.]

Figure 6.4.2   Lead flat roof details

COPPER-COVERED FLAT ROOFS

Copper, like lead, has been used as a building material for many centuries. It is obtained from ore deposits which have to be mined, crushed and ground to a fine powder. Generally by a system of flotation the copper dust, which is about 4% of the actual rock mined, is separated from the waste materials and transferred to the smelting furnace and then cast into cakes of blister copper or thick slabs called *anodes*. The metal is now refined and formed into copper wire, strip, sheets and castings. Copper is also used to form alloys which are used in making components for the building industry. Common alloys are copper and zinc (forming brass) and copper and tin (forming bronze).

Copper is a dense material ($8,900 \text{ kg/m}^3$) which is highly ductile and malleable and can be cold worked into the required shape or profile. The metal hardens with cold working but its original dead soft temper can be restored by the application of heat with a blow-lamp or oxy-acetylene torch and quenching with water or by natural air cooling. If the dead soft temper is not maintained the hardened copper will be difficult to work and may fracture. On exposure to the atmosphere copper forms on its upper surface a natural protective film or patina, which varies in colour from green to black, making the copper beneath virtually inert.

For covering flat roofs rolled copper complying with the recommendations of BS 2870 is generally specified. Rolled copper is available in three forms:

1. **Sheet** – flat material of exact length, over 0.15 mm up to and including 10.0 mm thick and over 450 mm in width.
2. **Strip** – material over 0.15 mm up to and including 10.0 mm thick and any width, and generally not cut to length. It is usually supplied in coils but can be obtained flat or folded.
3. **Foil** – material 0.15 mm thick and under, of any width supplied flat or in a coil; because of its thickness foil has no practical application in the context of roof coverings.

In general copper strip is used for flashings and damp-proof courses, whereas sheet or strip can be considered for general roof covering application; widths of 600 and 750 mm, according to thickness, are used with a standard length of 1.800 m.

Forming double lock welts

Single welt

Fall

Locked

Double welt

25

40

Upstands

Copper apron dressed into gutter

Fascia

Ventilated soffit

Vapour-check lined plasterboard

Double lock welts

Batten forming drip

Batten rolls at 500 to 650 c/c

3,000 max.

Fall

Copper clips at 750 c/c

Copper flashing turned into brick joint

150 high skirting

Firring

Joist

Decking of plywood or similar material

Felt underlay

Figure 6.5.1 Typical layout of copper flat roof

The copper sheet or strip is laid in bays between rolls or standing seams to allow for the expansion of the covering. Standing seams are not recommended for pitches under 5° since they may be flattened by foot traffic and become a source of moisture penetration due to a capillary action. The recommended maximum bay widths for the common sheet thicknesses used are:

- 0.60 mm thick – maximum bay width 600 mm
- 0.70 mm thick – maximum bay width 750 mm

Transverse joints in the form of double lock welts should be used on all flat roofs. The welts should be sealed with a mastic or linseed oil before being folded together. Cross-welts may be staggered or continuous across the roof when used in conjunction with batten rolls (see Fig. 6.5.1).

An exception is the relatively new concept of long strip copper roofing. It uses special sheet lengths of up to 10 m in standard widths, without the need for cross-welts. Panels are machined together with standing welts/seams which include sliding clips to allow for longitudinal movement. The process is less labour intensive than traditional, therefore more economical to apply and the roof appearance is smoother.

The substructure supporting the covering needs to be permanent, even, smooth and laid to the correct fall of not less than 1:60 (40 mm in 2.400 m). With a concrete structural roof a lightweight screed is normally laid to provide the correct surface and fall; if the screed has a sulphate content it will require sealing coats of bitumen to prevent any moisture present forming dilute acids which may react with the copper covering. The 50 × 25 dovetail battens inserted in the screed to provide the fixing medium for the batten rolls should be impregnated with a suitable wood preservative. Timber flat roofs should be wind tight and free from 'spring' and may be covered with 25 mm nominal tongued and grooved boarding laid with the fall or diagonally, or any other decking material which fulfils the general requirements given above.

On flat roofs drips at least 65 mm high are required at centres not exceeding 3.000 m to increase the flow of rainwater across the roof to the gutter or outlet. To lessen the wear on the copper covering as it expands and contracts a separating layer of felt (BS 747 Type 4A(ii)) should be incorporated into the design. The copper sheet should be secured with copper wire nails not less than 2.6 mm thick and at least 25 mm long; the batten rolls should be secured with well countersunk brass screws. Typical constructional details are shown in Fig. 6.5.2.

[*Note*: For habitable accommodation current building regulations will require the external walls to include a lightweight concrete block inner leaf and cavity insulation. The roof will also need insulating as shown in Chapter 6.1.]

Figure 6.5.2 Copper flat roof details

ROOFLIGHTS IN PITCHED ROOFS

Rooflights can be included in the design of a pitched roof to provide daylight and ventilation to rooms within the roof space or to supplement the daylight from windows in the walls of medium and large span single storey buildings.

In domestic work a rooflight generally takes one of two forms – the dormer window or skylight. A dormer window has a vertical sash and therefore projects from the main roof, the cheeks or sides can be clad with a sheet material such as lead or tile hanging and the roof can be pitched or flat of traditional construction (see Fig. 6.6.1).

A skylight is fixed in a trimmed opening and follows approximately the pitch of the roof. It can be constructed as an opening or dead light (see Fig. 6.6.2). In common with all rooflights in pitched roofs, making the junctions water- and weathertight present the greatest problems, and careful attention must be given to the detail and workmanship involved in the construction of dormer windows and rooflights.

Roofs of the type used on medium span industrial buildings with coverings such as corrugated fibre cement sheeting supported by purlins and steel roof trusses require a different treatment. The amount of useful daylight entering the interior of such a building from windows fixed in the external walls will depend upon:

- The size of windows.
- The height of window above the floor level.
- The span of building.

Generally the maximum distance useful daylight will penetrate is approximately 10.000 m; over this distance artificial lights or roof lights will be required during the daylight period. Three methods are available for the provision of roof lights in profiled sheeted roofs.

Special rooflight units of fibre cement consisting of an upstand kerb surmounted by either a fixed or opening glazed sash can be fixed instead of a standard profiled sheet. These units are useful where the design calls for a series of small isolated glazed rooflights to supplement the natural daylight. An alternative is to use translucent profiled sheets which are of the same size and profile as the main roof covering. In selecting the type of sheet to be used the requirements of Part B of

Timber studding and plasterboard to wall

Vapour check lined plasterboard

Dormer window

Horizontal section through cheek

Trimming rafter

Lead or similar cladding to dormer cheek (tiled or other vapour permeable treatment preferred)

Plywood face

Studding to cheek

Timber facing

Timber beam over dormer opening

Dormer cheeks of timber studding

Dormer wall of timber studding

Skirting

Flooring and joists

Insulated timber studding

Tilting fillet

Sheet roof covering

Roof decking

Wrot timber beam with outside timber facing

Vapour check lined plaster board

Roof joists

Firring

Fillet

Fascia

Ventilation gap

Soffit

Casement window

Lead or similar flashing

Roof tiling

Vertical section

Figure 6.6.1 Typical flat roof dormer window details

Figure 6.6.2 Typical timber opening skylight

the Building Regulations must be considered. Approved Document B4 (Section 14) deals specifically with the fire risks of roof coverings and refers to the designations defined in BS 476: Part 3. These designations consist of two letters, the first letter represents the time of penetration when subjected to external fire and the second letter is the distance of spread of flame along the external surface. Each group of designations has four letters A–D and in both cases the letter A has the highest resistance. Specimens used in the BS 476 test are either tested for use on a flat surface or a sloping surface and therefore the material designation is preceded by either EXT. F or EXT. S.

Most of the translucent profiled sheets have a high light transmission, are light in weight and can be fixed in the same manner as the general roof covering. It is advisable to weather seal all lapping edges of profiled rooflights with silicon or mastic sealant to accommodate the variations in thickness and expansion rate of the adjacent materials. Typical examples are:

Polyester glass fibre sheets　Made from polyester resins reinforced with glass fibre and nylon to the recommendations of BS 4154. These sheets can be of natural colour or tinted and are made to suit most corrugated fibre cement and metal profiles. Typical designations are EXT. S.AA for self-extinguishing sheets and EXT. S.AB for general-purpose sheets.

Wire-reinforced PVC sheets　Made from unplasticized PVC reinforced with a fine wire mesh to give a high resistance to shattering by impact. Designation is EXT. S.AA and they can therefore be used for all roofing applications. Profiles are generally limited to Categories A and B defined in BS EN 494.

PVC sheets　Made from heavy gauge clear unplasticized rigid PVC to the recommendations of BS 4203, are classified as self-extinguishing when tested in accordance with method 508A of BS 2782: Part O, Annex C, and may be used on the roof of a building provided that part of the roof is at least 6.000 m from any boundary. If that part of the roof is less than 6.000 m from any boundary and covers a garage, conservatory or out-house with a floor area of less than 40 m², or is on an open balcony, carport or the like, or a detached swimming pool, PVC sheets can be used without restriction. Table 6.3 of Approved Document B defines the use of these sheets for the roof covering of a canopy over a balcony, veranda, open carport, covered way or detached swimming pool.

As an alternative to profiled rooflights in isolated areas continuous rooflights can be incorporated into a corrugated or similar roof covering by using flat-wired glass and patent glazing bars. The bars are fixed to the purlins and spaced at 600 mm centres to carry either single or double glazing. The bars are available as a steel bar sheathed in lead or PVC or as an aluminium alloy extrusion (see Fig. 6.6.3). Many sections with different glass-securing techniques are manufactured under the patents granted to the producers but all have the same basic principles. The bar is generally an inverted 'T' section, the flange providing the bearing for the glass and the stem depth giving the bar its spanning properties. Other standard components

Figure 6.6.3 Typical patent glazing bar sections

are fixing shoes, glass weathering springs or clips and glass stops at the bottom end of the bar (see Fig. 6.6.3).

Since the glass and the glazing bar are straight they cannot simply replace a standard profiled sheet; they must be fixed below the general covering at the upper

Figure 6.6.4 Patent glazing and profiled roof covering connection details

Insulation

Steel roof truss rafter

Shelf angle bearer

Patent glazing bar fixing shoe

Angle cleat

Profiled roof sheeting

Eaves filler piece

Lead flashing

Patent glazing bar

Glass

Bolt and clip

Angle purlin

Patent glazing bar

Glass

Timber draught excluder with rubber seal and fixing shoe

Glass stop

Flashing piece

Shelf angle bearer

Steel plate batten strip

Steel roof truss rafter

Angle purlin

Bolt and clip

Angle cleat

Profiled roof sheeting

Figure 6.6.5   'Glidevale' patent double-glazing unit

end and above the covering at the lower end to enable the rainwater to discharge on to the general roof surface. Great care must be taken with this detail and with the quality of workmanship on site if a durable and satisfactory junction is to be made. Typical details are shown in Fig. 6.6.4.

The total amount of glazing to be used in any situation involves design appreciation beyond the scope of this volume, but a common rule of thumb method is to use an area of glazing equal to 10% of the total roof area. The glass specified is usually a wired glass of suitable thickness for the area of pane being used. Wired glass is selected so that it will give the best protection should an outbreak of fire occur; the splinters caused by the heat cracking the glass will adhere to the wire mesh and not shatter on to the floor below. A successful alternative is laminated glass composed of two or more sheets of float glass with a plastic interlayer. Polyvinylbutyral (PVB) is the most common interlayer, which is bonded to the glass laminates by heat and pressure. If broken, the glass remains stuck to the PVB to provide the benefit of safety and security. It also remains intact following impact from bullets or explosive blasts.

As with timber skylights, provision should be made to collect the condensation which can occur on the underside of the glazing to prevent the annoyance of droplets of water falling to the floor below. Most patent glazing bars for single glazing have condensation channels attached to the edges of the flange which directs the collected condensation to the upper surface of the roof below the glazing line (see Figs 6.6.3 and 6.6.4). Contemporary frames can be produced from uPVC or extruded aluminium. The latter is often coated in polyester powder for colour enhancement and protection. A double-glazed example is shown in Fig. 6.6.5. This applies the principle of tension between the glazing frame and glass for security of fit, with the ability to respond to thermal movement without loss of seal.

Polycarbonate (PC) can be an economic option in preference to glass, particularly for roof light applications. It has the benefit of a double skin if required and, although produced in flat sheet format, can be moulded to domed profiles. Colour variants are possible, in addition to transparent, translucent and opaque composition. Dimensions, types and characteristics are defined in BS EN ISO 11963.

PART 7

INTERNAL FIXTURES AND FITTINGS

DOORS, DOOR FRAMES AND LININGS

DOORS

A door is a screen used to seal an opening into a building or between rooms within a building. It can be made of timber, glass, metal or plastic or any combination of these materials. Doors can be designed to swing from one edge, slide, slide and fold or roll to close an opening. The doors to be considered in this book are those made of timber and those made of timber and glass which are hung so that they swing from one edge. All doors may be classified by their position in a building, by their function or by their method of construction.

EXTERNAL DOORS

These are used to close the access to the interior of a building and provide a measure of security. They need to be weather resistant since in general they are exposed to the elements; this resistance is provided by the thickness, stability and durability of the construction and materials used together with protective coatings of paint or polish. The external walls of a building are designed to give the interior of a building a degree of thermal and sound insulation; doors in such walls should therefore be constructed, as far as practicable, to maintain the insulation properties of the external enclosure.

The standard sizes for external timber doors are 1,981 mm high × 762 or 838 mm wide × 45 mm thick which is a metric conversion of the old Imperial door sizes. Metric doors are produced so that, together with the frame, they fit into a modular co-ordinated opening size and are usually supplied as door sets with the door already attached or hung in the frame.

INTERNAL DOORS

These are used to close the access through internal walls, partitions and to the inside of cupboards. As with external doors the aim of the design should be to maintain the properties of the wall in which they are housed. Generally internal doors are thinner than their external counterparts since weather protection is no longer a requirement. Standard sizes are similar to external doors but with a wider range of widths to cater for narrow cupboard openings.

Purpose-made doors

The design and construction of these doors is usually the same as manufacturers' standard doors but made to non-standard sizes, shapes or designs. Most door manufacturers produce a range of non-standard doors which are often ornate and are used mainly for the front elevation doors of domestic buildings. Purpose-made doors are also used in buildings such as banks, civic buildings, shops, theatres and hotels to blend with or emphasize the external façade design or internal decor (see Fig. 7.1.1).

Methods of construction

The British Standard Code of Practice 151 covers all aspects of door construction. Unfortunately it is no longer published, but access to old copies will provide a useful reference for all doors of timber construction and their frames. British Standard 459 is published especially for matchboarded wooden doors for use externally.

Standard doors are used extensively since they are mass produced to known requirements, are readily available from stock and are cheaper than purpose-made doors.

Panelled and glazed wood doors

The Code of Practice gives a wide variety of types all of which are based upon the one, two, three or four panel format. They are constructed of timber which should be in accordance with the recommendations of BS 1186 and BS EN 942 with plywood or glass panels. External doors with panels of plywood should be constructed using an external quality plywood (see Fig. 7.1.2).

The joints used in framing the doors can be a dowelled joint or a mortise and tenon joint. The dowelled joint is considered superior to the mortise and tenon joint and is cheaper when used in the mass production of standard doors. Bottom and lock rails have three dowels, top rails have two dowels and intermediate rails have a single dowel connection (see Fig. 7.1.3). The plywood panels are framed into grooves with closely fitting sides with a movement allowance within the depth of the groove of 2 mm. The mouldings at the rail intersections are scribed, whereas the loose glazing beads are mitred. Weatherboards for use on external doors can be supplied to fit onto the bottom rail of the door which can also be rebated to close over a water bar (see Fig. 7.1.2).

A fire door specification of FD 30 (see section headed: Flush fire doors) can be attained for 45 mm panelled doors by lining the existing panel with 6 mm glass-reinforced plasterboard or by replacing the panels with a double layer of 6 mm glass-reinforced plasterboard.

Horn

Top rail
Glass panel
Glazing bars
Frieze rail
Solid panel
Lock or middle rail
Stile
Intermediate rail
Muntin
Bottom rail

Door terminology

Stile or rail
Square sunk or flat panel
Ovolo 'on solid' or stuck moulding

Stile or rail
Bead and butt panel

Stile or rail
Framed moulding
Solid panel

Stile or rail
Planted moulding
Bevelled, raised and fielded panel
Bolection moulding
Bevel
Field
Raising

Stile or rail
Planted moulding
Slotted and screwed
Bolection moulding

Figure 7.1.1 Purpose-made doors and mouldings

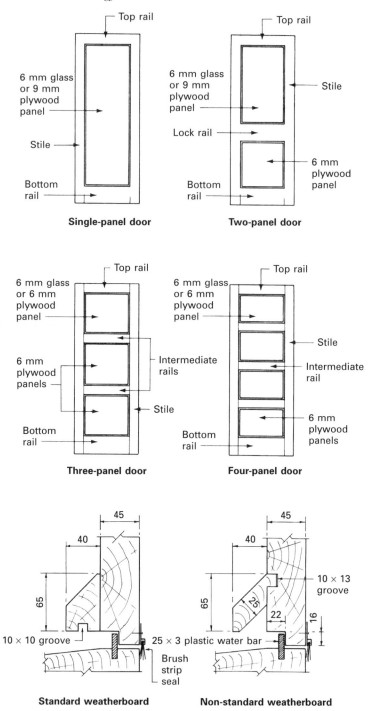

Figure 7.1.2   Standard panelled doors and weatherboards

35 or 45

95 deep top rail

Top rail

Scribed shoulder

Stile

16 mm dia. keyed dowels stopped 25 mm from outside edge of stile

Typical dowelled joint

Glass panel bedded in putty or leather

Glazing bead pinned to rail

95 × 35 or 45 intermediate rail

6 or 9 mm plywood panel

Haunch

Scribed shoulder

Top rail

Tenon

Stile

Mortice

Wedges

Typical mortice and tenon joint

195 × 35 or 45 bottom rail

Figure 7.1.3 Panelled door details

Flush doors

This type of door has the advantage of a plain face which is easy to clean and decorate; it is also free of the mouldings which collect dust. Flush doors can be faced with hardboard, plywood or a plastic laminate and by using a thin sheet veneer of good-quality timber the appearance of high-class joinery can be created.

The method of constructing flush doors is left to the manufacturer, which provides complete freedom for design. Therefore the forms of flush door construction are many and varied but basically they can be considered as either skeleton core doors or solid core doors. The former consists of an outer frame with small section intermediate members over which is fixed the facing material. The facing has a tendency to deflect between the core members and this can be very noticeable on the surface especially if the facing is coated with gloss paint. Solid core doors are designed to overcome this problem and at the same time improve the sound insulation properties of flush doors. They also provide a higher degree of fire resistance and security. Solid doors of suitably faced block or lamin board are available for internal and external use. Another method of construction is to infill the voids created by a skeleton core with a lightweight material such as foamed plastic which will give support to the facings but will not add appreciably to the weight of the door.

The facings of flush doors are very vulnerable to damage at the edges, therefore a lipping of solid material should be fixed to at least the vertical edges (good class doors have lippings on all four edges).

Small glazed observation panels can be incorporated in flush doors when the glass panel is secured by loose fixing beads (see Figs 7.1.4 and 7.1.5).

Flush fire doors

These doors provide an effective barrier to the passage of fire for the time designated by a reference coding. They can be assessed in terms of stability and integrity in the presence of fire tests. For example, a coding of 30/30 indicates the time in minutes a door is 'stable' under simulated fire conditions and sufficiently 'integral' to resist the penetration of flame and hot gases. To achieve this they must be used in conjunction with the correct frame.

However, the term *stability* is now considered less appropriate and in accordance with BS 476: Part 22: *Methods for determination of the fire resistance of non-loadbearing elements of construction*, integrity and insulation are sufficient criteria for grading (see BS 476: Part 20 for test procedures). Furthermore, BS 5588: *Fire precautions in the design, construction and use of buildings*, and the Building Regulations Approved Document B (see Appendix B, Table B1) designate fire doors by integrity performance only.

The performance of fire doors is simplified by expressing the category with the initials FD and the integrity in minutes numerically. For example, FD 60 indicates a fire door with 60 minute integrity, and if the coding is followed by the letter S it indicates the door or frame has a facility to resist the passage of smoke, i.e. FD 60S.

Figure 7.1.4 Skeleton core flush doors

Standard ratings are 20, 30, 60 and 90, but other ratings are available to order. A timber frame is unlikely to satisfy British Standard tests for a 120 rating.

Resistance to smoke penetration can be achieved by fitting a brush strip seal to the frame. These are primarily for draught-proofing, but will be effective in providing a barrier to cold smoke. A more effective barrier to smoke and flame is

Figure 7.1.5   Solid core doors

an intumescent strip fitted to the door edge or integral with the frame. The latter is usually the most practical, being secured in a frame recess with a rubber based or PVA adhesive. At temperatures of about 150 °C the seal expands to prevent passage of fire and smoke for the designated period. The seal will not prevent door movement for access and escape of personnel.

Notes

1. Doorsets/hardware and glass must be selected with compatibility to the door specification, (see BS 476: Part 31 for conformity of doorsets).
2. A colour-coded plastic plug is fitted into the hanging style of the door to indicate fire resistance. For example:

   - Blue core/white background:
     - FD 20 without intumescent strip in frame.
     - FD 30 with intumescent strip in frame.
   - Red core/blue background:
     - FD 60 with two intumescent strips in frame.

   A full list of coloured core and background codes can be found in BS 8214, the Code of practice for fire door assemblies (see Table 1).
3. Half hour fire rated doors are hung using one pair of hinges, whereas one hour doors require one and a half pairs of hinges (see Fig. 7.1.6).
4. All fire doors should have an automatic self-closing device, excepting cupboards and accesses to service ducts which are normally kept locked.

Matchboarded doors

These doors can be used as external or internal doors and as a standard door takes one of two forms, a ledged and braced door or a framed ledged and braced door, the latter being a stronger and more attractive version.

The face of the door is made from tongue and grooved boarding which has edge chamfers to one or both faces; these form a vee joint between consecutive boards. Three horizontal members called *ledges* clamp the boards together and in this form a non-standard door, called a ledged and battened door, has been made. It is simple and cheap to construct but has the disadvantage of being able to drop at the closing edge thus pulling the door out of square – the only resistance offered is that of the nails holding the boards to the ledges. The use of this type of door is limited to buildings such as sheds, outhouses and to small units like trapdoors (see Fig. 7.1.7). In the standard door braces are added to resist the tendency to drop out of square; the braces are fixed between the ledges so that they are parallel to one another and slope downwards towards the hanging edge (see Fig. 7.1.7).

In the second standard type a mortise and tenoned frame surrounds the matchboarded panel giving the door added strength and rigidity (see Fig. 7.1.7). If wide doors of this form are required the angle of the braces becomes too low to be of value as an effective restraint and the brace must therefore be framed as a diagonal between the top and bottom rails. Wide doors of this design are not covered by the British Standard but are often used in pairs as garage doors or as wide entrance doors to workshops and similar buildings (see Fig. 7.1.8).

The operation of fixing a door to its frame or lining is termed *hanging* and entails removing the protective horns from the top and bottom of the stiles; planing the

838 or 914

83 × 38 framing

Lipping

45 mm deep rail

165 mm deep rail

9.5 mm plasterboard or 6 mm glass-reinforced plasterboard protective lining to both faces

3 mm plywood facing

Half-hour fire door and frame (FD 30)

59

25 × 35 planted stop

83

Intumescent strip

45

34

3 mm max. gap

Two-piece frame

838 or 914

83 × 38 framing

Lipping

45 mm deep rail

165 mm deep rail

9.5 mm plasterboard or 6 mm glass-reinforced plasterboard protective lining to both faces

6 mm cement particle board to both faces

3 mm plywood facing

One hour fire door and frame (FD 60)

59

35

Dual intumescent strip

92

54

34

3 mm max. gap

Solid frame

Figure 7.1.6   Fire doors and frames

stiles to reduce the door to a suitable width; cutting and planing the top and bottom to the desired height; marking out and fitting the butts or hinges which attach the door to the frame and fitting any locks and door furniture which is required. The hinges should be positioned 225 mm from the top and bottom of the door and

16 mm match boarding
150
95 × 22 ledge
95 × 22 brace
95 × 22 ledge
Hanging edge
95 × 22 brace
150
95 × 22 ledge

Ledged and battened
(rear elevation)

Ledged and braced
(rear elevation)

16 mm match boarding
95 × 45 framing
95 × 27 brace
Hanging stile
146 × 27 middle rail
95 × 27 brace
146 × 27 bottom rail

25
15
Brace
Ledge
15

Alternative brace connection

t
$5t$ max.
Bottom rail

Barefaced tenon

Framed, ledged and braced
(rear elevation)

Figure 7.1.7   Matchboarded doors

Hanging stile

Meeting stile

95 × 45 framing

16 mm matchboarding

146 × 27 brace

197 × 27 bottom and middle rails

Rear elevation

45 × 12 cover fillet

Meeting stile

Matchboarding

15

Rebated
meeting stile

Matchboarding

Alternative meeting stile treatments

45

Rebated stile

Matchboarding

Alternative stile treatment

Figure 7.1.8   Matchboarded double doors

where one and a half pairs of hinges are specified for heavy doors the third hinge is positioned midway between the bottom and top hinges.

A door, irrespective of the soundness of its construction, will deteriorate if improperly treated during transportation, storage and after hanging. It should receive a wood priming coat of paint before or immediately after delivery, and be stored in the dry and in a flat position so that it does not twist – it should also receive the finishing coats of paint as soon as practicable after hanging.

FRAMES AND LININGS

A door frame or lining is attached to the opening in which a door is to be fitted, it provides a surround for the door and is the member to which a door is fixed or hung. Door sets consisting of a storey height frame with a solid or glazed panel over the door head are also available; these come supplied with the door ready hung on lift-off hinges (see Fig. 7.1.9).

TIMBER DOOR FRAMES

These are made from rectangular section timber in which a rebate is formed or to which a planted door stop is fixed to provide the housing for the door. Generally a door frame is approximately twice as wide as its thickness plus the stop. Frames are used for most external doors, heavy doors, doors situated in thin non-loadbearing partitions and for internal fire doors.

A timber door frame consists of three or four members – namely the head, two posts or jambs and a sill or threshold. The members can be joined together by wedged mortise and tenon joints, combed joints or mortise and tenon joints pinned with a metal star-shaped dowel or a round timber dowel. All joints should have a coating of adhesive (see Fig. 7.1.10).

Door frames which do not have a sill are fitted with mild steel dowels driven into the base of the jambs and cast into the floor slab or, alternatively, grouted into preformed pockets as a means of securing the feet of the frame to the floor. If the frame is in an exposed position it is advisable to sit the feet of the jambs on a damp-proof pad such as lead or bituminous felt, to prevent moisture soaking into the frame and creating the conditions for fungi attack.

Door frames fitted with a sill are designed for one of two conditions:

1. Doors opening out.
2. Doors opening in.

In both cases the sill must be designed to prevent the entry of rain and wind under the bottom edge of the door. Doors opening out close onto a rebate in the sill whereas doors opening in have a rebated bottom rail and close over a water bar set into the sill (see Fig. 7.1.10).

Typical door lining

Typical door set

Figure 7.1.9   Door linings and door sets

Horn built in or cut off

Mortice

85 × 57 splayed and rebated head

10 mm dia. wood dowel

Tenon full width of jamb

3 No. corrugated tee brackets per jamb

16

25.5

51

52 48

100

1.5 mm galvanized mild steel

External metal door frame

Metal fixing cramp screwed to back of frame

85 × 57 splayed and rebated jamb

13 mm deep rebate for door

10 mm dia. wood dowel

Tenon

Hole for dowel

25 × 3 plastic water bar

Mortice

10 mm throat

25 mm wide bedding groove

121 × 45 hardwood weathered sill

Figure 7.1.10 Door frames

Timber door frames can be fixed to a wall by the following methods:

1. Built into the brick or block wall as the work proceeds by using L-shaped ties or cramps. The ties are made from galvanized mild steel with one end turned up 50 mm, with two holes for wood screws, the other end being 100 to 225 mm long and fish-tailed for building into brick or block bed joints. The ties are fixed to the back of the frame for building in at 450 mm centres.
2. Fixed into a brick opening at a late stage in the contract to prevent damage to the frame during the construction period. This is a more expensive method and is usually employed only when high-class joinery using good-quality timber is involved. The frames are fixed to timber plugs inserted into the reveals with wood screws whose heads are sunk below the upper surface of the frame; this is made good by inserting over the screw heads plugs or pellets of matching timber.

Timber door frames of softwood are usually finished with several applications of paint, whereas frames of hardwood are either polished or oiled. Frames with a factory coating of plastic are also available.

METAL DOOR FRAMES

These are made from mild steel pressed into one of three standard profiles and are suitable for both internal and external positions. The hinges and striking plates are welded on during manufacture and the whole frame receives a rust-proof treatment before delivery. The frames are fixed in a similar manner to timber frames using a tie or tee bracket which fits into the back of the frame profile and is built into the bed joints of the wall (see Fig. 7.1.10). The advantage of this type of frame is that they will not shrink or warp but they are more expensive than their timber counterparts.

DOOR LININGS

These are made from timber board 25 or 32 mm thick and as wide as the wall plus any wall finishes. They are usually only specified for internal doors. Door linings are not built in but are fixed into an opening by nailing or screwing directly into block walls or into plugs in the case of brick walls. Timber packing pieces or folding wedges are used to straighten and plumb up the sides or jambs of the lining (see Fig. 7.1.9).

GLASS AND GLAZING

GLASS

Glass is made mainly from soda, lime, silica and other minor ingredients such as magnesia and alumina, to produce a material suitable for general window glazing. The materials are heated in a furnace to a temperature range of 1,490–1,550 °C when they fuse together in a molten state – they are then formed into sheets by a process of drawing, floating or rolling. BS 952: Part 1: *Glass for glazing*, classifies glass for use in buildings by composition, dimensions, mass and available sizes.

DRAWN CLEAR SHEET GLASS

There are two principal methods of producing drawn clear sheet glass: the first is by vertical drawing from a pool of molten glass which, when 1 m or so above the pool level, is rigid enough to be engaged by a series of asbestos faced rollers that continue to draw the ribbon of glass up a tower some 10 m high, after which the ribbon is cut into sheets and washed in a dilute acid to remove surface deposits. In the second method the glass is initially drawn in the vertical plane but it is turned over a roller so that it is drawn in the horizontal direction for some 60 m and passes into an annealing furnace, at the cold end of which it is cut into sheets.

Clear sheet glass is a transparent glass with 85% light transmission, with a fire-finished surface, but because the two surfaces are never perfectly flat nor parallel there is always some distortion of vision and reflection.

The only thicknesses produced in the UK are 3 and 4 mm; 5 and 6 mm are available as an import.

FLOAT GLASS

This is gradually superseding the sheet glass process and provides a glass with undistorted vision. It is formed by floating a continuous ribbon of molten glass over a bath of liquid metal at a controlled rate and temperature. A general glazing quality and a selected quality are produced in thicknesses ranging from 3 to 25 mm.

ROLLED AND ROUGH CAST GLASS

This is a term applied to a flat glass produced by a rolling process. Generally the glass produced in this manner is translucent, which transmits light with varying degrees of diffusion so that vision is not clear. A wired transparent glass with 80% light transmission is, however, produced generally in one thickness of 6 mm. The glass is made translucent by rolling on to one face a texture or pattern which will give 70–85% light transmission. Rough cast glass has an irregular texture to one side; wired rough cast glass comes in two forms, Georgian wired, which has a 12 mm square mesh electrically welded wire reinforcement, or hexagonally wired which is reinforced with hexagonal wire of approximately 20 mm mesh. Rough cast glass is produced in 5, 6 and 10 mm thicknesses and is made for safety and fire resistant glazing purposes.

SPECIAL GLASS PRODUCTS

Numerous special glasses are available, some of the most significant include:

- Toughened glass.
- Laminated glass.
- Fire-resisting glass.

Toughened glass

Processed glass is subjected to heat and then cooled by jets of cold air. The effect is to create compressive stresses at the surface with balancing tensile stresses in the centre. The result is a glass of a strength 4 to 5 times greater than annealed glass. It can, however, still fracture under extreme bending loads or severe impact from a sharp implement. When broken it shatters into small blunt-edged fragments which reduce the risk of personal injury. Due to its great strength and potential for large glazed areas, it provides considerable scope for freedom of design.

Laminated glass

As the name suggests, laminated glass is built up in layers. It is composed of an outer and inner layer of annealed float glass with a heat and pressure-sealed intermediate bonding layer of polyvinylbutyral (PVB). Performance can be varied by changing the number and thickness of glass layers to provide a wide range of applications and purposes. These include rooflights, internal doors and partitions, noise and solar controls as well as its suitability for security and safety situations due to its inherent resistance to bullets and explosives.

Fire-resisting glass

Fire resistance can be measured in terms of *integrity* and *insulation* – see also Chapter 7.1. BS 476: Part 20, defines integrity as, 'The ability of a specimen of

Table 7.2.1   Fire test performance of Pilkington Pyrodur and Pyrostop glass

Glass type	Thickness (mm)	Integrity (min.)	Insulation (min.)
Pyrodur	10	30	N/A
	13	60	N/A
Pyrostop	15	60	30
	21	60	60
	44	90	90
	50	120	120

a separating element to contain a fire to specified criteria of collapse, freedom from holes, cracks and fissures and sustained flaming on the unexposed face'. Insulation is also defined in the same BS as, 'The ability of a specimen of a separating element to restrict the temperature rise of the unexposed face to below specified levels'. Specified levels means an average of no more than 140 °C and in any specific position 180 °C.

Wired glass, composed of an electrically welded mesh embedded centrally in rough glass during the rolling process, is well established as an effective means of resisting fire. It is frequently used in doors for security reasons and its ability to remain integral in fire for up to two hours. Although the glass fractures in heat, the close proximity of mesh retains the unit in one piece. Clear fire-resisting glass is produced by Pilkington UK Ltd under the trade names or Pyrodur and Pyrostop. Pyrodur is suitable for use internally or externally in doors and screens. Standard thicknesses are 10 and 13 mm, composed of three or four glass layers, respectively. Interlayers are of intumescent material with an ultraviolet bonding laminate of polyvinylbutyral to ensure impact resistance performance. It is primarily intended to satisfy the Building Regulations' requirements for integrity, as shown in Table 7.2.1, although it has insulation properties for up to 22 minutes. Pyrostop is produced in two grades corresponding to internal or external application. For internal use it is made up of four glass layers with three intumescent interlayers. External specification has an additional glass and an ultraviolet filter interlayer positioned to the outside to protect the intumescent material against degradation. It is also produced in double-glazed format with an 8 mm air gap achieved with steel spacer bars. Overall thicknesses are 15, 21, 44 and 50 mm, to attain integrity and insulation for up to two hours.

GLAZING

The securing of glass in prepared openings such as doors, windows and partitions is termed *glazing*. It is logical that as the area of glass pane increases so must its thickness; similarly, position, wind load and building usage must be taken into account.

Figure 7.2.1   Zones of safe glazing – shown shaded

The Building Regulations effect suitable standards for the safe use and application of glass through Approved Document N. Position is very important and critical locations are shown in Fig. 7.2.1. These are defined within 800 mm of the finished floor level and 1,500 mm if in a door or adjacent side panel. Glass in critical locations must be classified as safe, which means it satisfies one of the following:

1. It conforms to the 'safe break' characteristics defined in BS 6206: *Specification for impact performance requirements for flat safety glass and safety plastics for use in buildings.*
2. It is robust or limited to small panes. Robust glass is inherently strong, being manufactured from material such as polycarbonates. Annealed glass may be used within the limitations of Table 7.2.2. Small panes can be isolated or in groups separated by glazing bars. The maximum pane width is 250 mm and the maximum pane area is 0.5 m². Nominal thickness is at least 6 mm.
3. It is permanently protected by a screen, as shown in Fig. 7.2.2.

Table 7.2.2 Maximum dimensions of annealed glass panels

Annealed glass thickness (mm)	Maximum width (m)	Maximum height (m)
8	1.10	1.10
10	2.25	2.25
12	4.50	3.00
15	Any	Any

Figure 7.2.2 Screen protection to glazing

The effect of wind loading on glazed area and glass thickness can be determined by reference to BS 6262: *Code of practice for glazing for buildings*. Tables and graphs permutate data for topography, ground roughness, life and glass factors to provide the recommended minimum glass thickness.

GLAZING WITHOUT BEADS

This is a suitable method for general domestic window and door panes; the glass is bedded in a compound and secured with sprigs, pegs or clips and fronted with a weathered surface putty. Putty is a glazing compound which will require a protective coating of paint as soon as practicable after glazing. Two kinds of putty are in general use:

1. **Linseed oil putty**: for use with primed wood members and is made from linseed oil and whiting, usually to the recommendations of BS 544.

2. **Metal casement putty**: for use with metal or non-absorbent wood members and is made from refined vegetable drying oils and finely ground chalk.

[*Note*: A general-purpose putty is available which will satisfy application for both wood and metal glazing.]

The glass pane should be cut to allow a minimum clearance of 2 mm all round for both wood and metal frames. Sufficient putty is applied to the rebate to give at least 2 mm of back putty when the glass is pressed into the rebate, any surplus putty being stripped off level or at an angle above the rebate. The glass should be secured with sprigs or clips at not more than 440 mm centres and finished off on the front edge with a weathered putty fillet so that the top edge of the fillet is at or just below the sight line (see Fig. 7.2.3).

Compounds and sealants

Putty has the disadvantage of setting after a period of time, which may be unsuitable for large glazed areas subject to thermal movement. Therefore several non-setting compounds and sealants are available. Rubberized compounds are usually in two parts for mixing on site and applying by hand or with an extruder gun. One part sealants of polysulphide, silicone and urethane cure by chemical reaction with exposure to the atmosphere. They form a firm resilient seal with a degree of flexibility. One part sealants of butyl or acrylic do not cure, but remain soft and pliable. These are preferred for bedding glass and beads only, as they can deteriorate if exposed. Two parts polysulphide- or polyurethane-based sealants are mixed for application to cure into rubbery material within 14 days. These have excellent adhesion.

GLAZING WITH BEADS

For domestic work glazing with beads is generally applied to good class joinery. The beads should be secured with either panel pins or screws – for hardwoods it is usual to use cups and screws. The glass is bedded in a compound or a suitable glazing felt mainly to prevent damage by vibration to the glass. Beads are usually mitred at the corners to give continuity of any moulding. Beads for metal windows are usually supplied with the surround or frame and fixing of glass should follow the manufacturer's instructions (see Fig. 7.2.3).

DOUBLE GLAZING

Double glazing is now standard practice in window installation to satisfy UK Building Regulation requirements for energy conservation in external walls. Factory made units contain two parallel glass panes sealed with an air gap between the glass. This not only reduces heat loss considerably but will also improve sound insulation and restrict surface condensation. Typical manufactured units comprise two panes with a dry air gap of between 3 and 20 mm. The greater

12–20

Sprig

Back putty

Glass

Front putty

Sprig

Spring clip

Back putty

Wood frame

Metal frame

Glazing without beads

Glass

Internal bead fixed with cups and screws

External bead

Bedding material

Wood frame

Metal frame

Mastic or sealant recommended by manufacturer

Glazing with beads

Figure 7.2.3 Glazing details

the air space, the better the thermal insulation, although care must be observed when applied at extremes of atmospheric conditions as explosion or implosion may occur with wider gapped units. Generally, the insulative effect is to reduce the thermal transmittance coefficient (U value) by about half, from approximately 5.6 W/m^2 K for single glazing to approximately 2.8 W/m^2 K, depending on air gap and exposure. Some applications are shown in Fig. 7.2.4.

Figure 7.2.4   Double glazing

WINDOWS

The primary function of a window is to provide a means for admission of natural daylight to the interior of a building. A window can also serve as a means of providing the necessary ventilation of dwellings, as required under Building Regulation F1, by including into the window design opening lights.

Windows, like doors, can be made from a variety of materials or a combination of these materials such as timber, metal and plastic. They can also be designed to operate in various ways by arranging the sashes to slide, pivot or swing, or hang as a casement to one of the frame members.

BUILDING REGULATIONS

Approved Document F details the requirements for ventilating habitable rooms, kitchens, utility rooms, bathrooms and sanitary accommodation in domestic buildings. A habitable room is defined as a room used for dwelling purposes but not a kitchen. A habitable room must have ventilation openings unless it is adequately ventilated by mechanical means.

Passive stack ventilation (PSV) is an acceptable alternative in some rooms (see Table 1, Approved Document F). This is a simple system of ducts used to convey air by natural movement to external terminals. Non-domestic buildings have different requirements to suit various purposes, with criteria applied to occupied rooms instead of habitable rooms and greater emphasis on mechanical ventilation and floor areas (see Table 2, Approved Document F).

A ventilation opening includes any permanent or closable means of ventilation which opens directly to the external air. Included in this category are openable parts of windows, louvres, air bricks, window trickle ventilators and doors which open directly to the external air. The three general objectives of providing a means of ventilation are:

- To extract moisture from rooms where it is produced in significant quantities such as kitchens and bathrooms.
- To provide for occasional rapid ventilation where moisture presence is likely to produce condensation.

■ To achieve a background ventilation which is adequate without affecting the comfort of the room.

Approved Document F provides the following for domestic dwellings:

1. **Habitable rooms** – provision of ventilation opening(s) with a total area of at least one-twentieth of the floor area of the room or rooms it serves and with some part of the ventilation opening(s) at least 1,750 mm above the floor level. Background or trickle ventilation is to be provided (normally in the window frame) having a total area of not less than 8,000 mm² by a means which is controllable, secure and sited to avoid draughts.
2. **Kitchens** – provision of rapid ventilation by intermittent operation of mechanical extract ventilation rated at not less than 60 litres per second. If incorporated in or adjacent to a cooker hood to be rated at not less than 30 litres per second. Background trickle ventilation provision to be not less than 4,000 mm².
3. **Bathrooms** – ventilation requirements can be satisfied by using intermittent mechanical extract ventilation rated at not less than 15 litres per second and trickle ventilation of at least 4,000 mm².
4. **Sanitary accommodation** – ventilation requirements can be satisfied with an openable window equivalent to one-twentieth of the floor area or intermittent mechanical extract ventilation of at least 6 litres per second.
5. If kitchens, utility rooms, bathrooms or lavatories are without openable windows, the combined light/extract fan switch must contain a 15 minute over-run facility for the fan.

Current UK energy conservation measures require windows to be double glazed. Also, all window sashes and doors must be draught-proofed with frames mastic or silicone sealed to the adjacent masonry. For domestic dwellings, the maximum glazed area of windows and doors should not exceed 22.5% of the total floor area. However, some variation is permissible where average glazing U values deviate from 3.0 or 3.3 W/m² K, depending on the insulation category of the house (see Building Regulations, Approved Document L, Table 3). Other residential buildings can have up to 30% of the exposed wall in glazed area and skylights occupying up to 20% of the roof area.

TRADITIONAL CASEMENT WINDOWS

Figure 7.3.1 shows a typical arrangement and details of this type of window. A wide range of designs can be produced by using various combinations of the members, the only limiting factor being the size of glass pane relevant to its thickness.

The general arrangement of the framing is important; heads and sills always extend across the full width of the frame and in many cases have projecting horns for building into the wall. The jambs and mullions span between the head and sill; these are joined to them by a wedged or pinned mortise and tenon joint. This arrangement gives maximum strength since the vertical members will act as struts – it will also give a simple assembly process.

External elevation

Mullion section

Jamb section

Vertical section

Figure 7.3.1 Traditional timber casement window

The traditional casement window frame has deep rebates to accommodate the full thickness of the sash, which is the term used for the framing of the opening ventilator. If fixed glazing or lights are required it is necessary to have a sash frame surround to the glass since the depth of rebate in the window frame is too great for direct glazing to the frame.

Standard wood casement windows

BS 644, Part 1: *Specification for factory assembled windows of various types*, provides details of the quality, construction and design of a wide range of wood casement windows. The frames, sashes and ventlights are made from standard sections of softwood timbers arranged to give a variety in design and size. The sashes and ventlights are designed so that their edges rebate over the external face of the frame to form a double barrier to the entry of wind and rain. The general construction is similar to that described for traditional casement windows and the fixing of the frame into walls follows that described for door frames.

Most joinery manufacturers produced a range of modified standard casement windows following the basic principles set out in BS 644 but with improved head, sill and sash sections. The range produced is based on brick modules to avoid unnecessary cutting and disturbance to the masonry appearance. Window types are identified by a notation of figures and letters, for example

212 CV

where   2 = width divided into two units
       12 = 1,200 mm height
        C = casement
        V = ventlight

For typical details see Figs 7.3.1–7.3.3.

Steel casement windows

These windows are produced to conform to the recommendations of BS 6510 which gives details of construction, sections, sizes, composites and hardware. The standard range covers fixed lights, hung casements, pivot casements and doors. They are produced in dimensions to suit both imperial and metric applications, the latter conforming to the basic space first preference of 300 mm giving the following range (in mm):

500; 600; 800; 900; 1,200; 1,500; 1,800.

Frame heights are based upon basic spaces for the preferred head and sill heights for public sector housing giving the following sizes (in mm):

200; 500; 700; 900; 1,100; 1,300; 1,500.

Steel windows, like wood windows, are identified by a notation of numbers and letters:

- Prefix number: × 100 = basic space length.
- Code letters: F = fixed light.
                   C = side-hung casement opening out.
                   V = top hung casement opening out and extending full width of frame.

Horn

Mullion

Top hung ventlight

Casement sash

Head

900

V

C

C

Hinge

Fixed light

Transom

Sill

External elevation (309 CVC window)

95 × 57 head

70 × 57 mullion

10

Fixed light

Casement sash

Mullion section

41 × 46 ventlight framing

70 × 57 transom

12

Direct glazing to fixed light

70 × 57 jamb

41 × 46 casement sash framing

Jamb section

95 × 70 softwood or hardwood sill

Vertical section

Figure 7.3.2 Typical modified BS casement window (see Fig. 7.3.3 for double-glazed vertical section)

Figure 7.3.3   Double-glazed casement window

> T = top hung casement opening out and extending less than
>   full width of frame.
> B = bottom casement opening inwards.
> S = fixed sublight.
- Suffix number: × 100 = basic space height.
- Suffix code: RH = right-hand casement as viewed from outside.
>   LH = left-hand casement as viewed from outside.

   The basic range of steel windows can be coupled together to form composite frames by using transom and mullion coupling sections without increasing the basic space module of 100 mm. The actual size of a steel frame can be obtained by deducting the margin allowance of 6 mm from the basic space size.

   All the frames are made from basic rolled steel sections which are mitred and welded at the corners to form right-angles; the internal bars are tenoned and riveted to the outer frame and to each other. The completed frame receives a hot dip galvanized protective finish after manufacture and before delivery.

Steel windows can be fixed into an opening by a number of methods such as direct fixing to the structure or by using a wood surround which is built into the reveals and secured with fixing ties or cramps. The wood surround will add 100 or 50 mm to the basic space size in each direction using either a nominal 75×75 mm or 50×75 mm timber section. Typical details of steel windows, couplings and fixings are shown in Figs 7.3.4 and 7.3.5.

The main advantage of steel windows is the larger glass area obtained for any basic space size due to the smaller frame sections used. The main disadvantage is the condensation which can form on the frames because of the high conductivity of the metal members.

BAY WINDOWS

Any window which projects in front of the main wall line is considered to be a bay window; various names are, however, given to various plan layouts (see Fig. 7.3.6). Bay windows can be constructed of timber, and/or metal and designed with casement or sliding sashes; the main difference in detail is the corner post, which can be made from the solid, jointed or masked in the case of timber and tubular for metal windows (see Fig. 7.3.6).

The bay window can be applied to one floor only or continued over several storeys. Any roof treatment can be used to cover in the projection and weather seal it to the main wall (see Fig. 7.3.7). No minimum head-room heights for bay windows or habitable rooms is given in the Building Regulations but 2.000 in bay windows and 2.300 in rooms would be considered reasonable. A bay window which occurs only on upper storeys is generally called an oriel window.

DOUBLE-HUNG SASH WINDOWS

These windows are sometimes called vertical sliding sash windows and consist of two sashes sliding vertically over one another. They are costly to construct but are considered to be more stable than side-hung sashes and have a better control over the size of ventilation opening thus reducing the possibility of draughts.

In timber two methods of suspension are possible:

1. Weight balanced type.
2. Spring balanced type.

The former is the older method in which the counterbalance weights suspended by cords are housed in a boxed framed jamb or mullion and have been generally

Top hung ventlight

Casement sash

Hinge

Fixed light

Mullion

Frame

External elevation (18 FCT 11 RH window)

25 × 33 o/a rebated frame

25 × 45 o/a rebated mullion

25 × 33 o/a casement sash framing

Mullion section

25 × 33 o/a rebated ventlight framing

25 × 28 o/a rebated transom bar

25 × 33 o/a casement sash framing

25 × 33 o/a rebated frame

Jamb section

25 × 33 o/a rebated frame

Vertical section

Figure 7.3.4   Typical steel window details

Figure 7.3.5 Steel window couplings and fixings

Figure 7.3.6   Bay window types and corner posts

superseded by the metal spring balance which uses a solid frame and needs less maintenance (see Figs 7.3.8 and 7.3.9).

Double hung sashes in metal are supported and controlled by spring balances or by friction devices but the basic principles remain the same.

DPC and flashing

2-layer mineralized built-up roofing felt

Compressed straw slab decking

Firring

Timber fillet

Fascia

Ventilation gap

RC beam

Plasterboard

Roof joists

Bay window

DPC and flashing

Lead roll

Lead sheet covering

Felt underlay

Plywood decking

UB

Timber fillet

Fascia

Ventilation gap

Cradling

Firring

Joist

Plasterboard

Bay window

Studding to underside of sill

Wall insulation

Vertical tile hanging

Bay window

RC beam with cantilever joists over

Note: Current practice is to insulate flat roofs (see Chapter 6.1) and stud framing (see Chapter 3.5)

Figure 7.3.7 Typical existing bay window roofs

Figure 7.3.8   Traditional double-hung weight-balanced sliding sash windows

PIVOT WINDOWS

The basic construction of the frame and sash is similar to that of a standard casement frame and sash. The sash can be arranged to pivot horizontally or vertically on friction pivots housed in the jambs or in the sill and head. These windows give good adjustment for ventilation purposes and in the main both faces of the glazing can be cleaned from the inside of the building.

40 mm min. horn projection

Dowelled mortice and tenon joint

114 × 33 solid head

48 × 21 head outside lining

19 × 14 removable bead

21 × 8 parting bead

48 × 41 upper sash framing

21 × 28 parting bead

114 × 33 solid stile

19 × 14 removable bead

Spring balance

48 × 21 outside lining

38 × 24 splayed end rebated meeting rails

Fixing hole

Tube housing coil spring

Rotating nylon bush

Variable twisted rod

Bracket

'Unique' spiral balance

48 × 14 lower sash framing

33 × 19 draught stop

135 × 60 flush sill

57 × 41 bottom rail

Note: If 114 × 60 solid stiles are used, balances can be housed in grooves within the stile thickness

Figure 7.3.9 Double-hung spring-balanced sash windows

TIMBER STAIRS

A stair is a means of providing access from one floor level to another. Modern stairs with their handrails are designed with the main emphasis on simplicity, trying to avoid the elaborate and costly features used in the past.

The total rise of a stair, that is, the distance from floor finish to floor finish in any one storey height, is fixed by the storey heights and floor finishes being used in the building design; therefore the stair designer only has the total going or total horizontal distance with which to vary the stair lay-out. It is good practice to keep door openings at least 450 mm away from the head or the foot of a stairway and to allow at least the stair width as circulation space at the head or foot of the stairway.

Stairs can be designed as one straight flight between floor levels which is the simplest and cheapest layout, alternatively they can be designed to turn corners by the introduction of quarter space (90°) or half space (180°) intermediate landings. Stairs which change direction of travel use tapered steps or are based on geometrical curves in plan. Irrespective of the plan lay-out the principles of stair construction remain constant and are best illustrated by studying the construction of simple straight flight stairs in this volume.

Terminology

- **Stairwell** The space in which the stairs and landings are housed.
- **Stairs** The actual means of ascension or descension from one level to another.
- **Tread** The upper surface of a step on which the foot is placed.
- **Nosing** The exposed edge of a tread, usually projecting with a square, rounded or splayed edge.
- **Riser** The vertical member between two consecutive treads.
- **Step** Riser plus tread.
- **Going** The horizontal distance between two consecutive risers or, as defined in Approved Document K, the horizontal dimensions from front to back of a tread less any overlap with the next tread above.
- **Rise** The vertical height between two consecutive treads.
- **Flight** A series of steps without a landing.
- **Newel** Post forming the junction of flights of stairs with landings or carrying the lower end of strings.

- **Strings** The members receiving the ends of steps which are generally housed to the string and secured by wedges, called wall or outer strings according to their position.
- **Handrail** Protecting member usually parallel to the string and spanning between newels. This could be attached to a wall above and parallel to a wall string.
- **Baluster** The vertical infill member between a string and handrail.
- **Pitch line** A line connecting the nosings of all treads in any one flight.

Building Regulations Part K

Approved Document K defines three categories of stairs:

1. **Private**: which is intended for use solely in connection with one dwelling.
2. **Institutional and assembly**: serves one place in which a substantial number of people will gather.
3. **Other**: which serves in buildings other than dwellings, institutional and assembly.

The practical limitations applicable to each of these three categories is shown in Table 7.4.1.

Table 7.4.1 Stairs (Approved Document K)

Category	Rise (mm)		Going (mm)
Private	155–220 or 165–200 with		245–260 or 223–300
Institutional and assembly*	135–180	with	280–340
Other*	150–190	with	250–320

* Subject to requirements for access for disabled people. See Building Regulations, Approved Document M2, Section 2.21 (max. rise 170 mm, min. going 250 mm).

CONSTRUCTION AND DESIGN

It is essential to keep the dimensions of the treads and risers constant throughout any flight of steps to reduce the risk of accidents by changing the rhythm of movement up or down the stairway. The height of the individual step rise is calculated by dividing the total rise by the chosen number of risers. The individual step going is chosen to suit the floor area available so that it, together with the rise, meets the requirements of the Building Regulations (see Fig. 7.4.1). It is important to note that in any one flight there will be one more riser than treads since the last tread is in fact the landing.

Stairs are constructed by joining the steps into the spanning members or strings by using housing joints, glueing and wedging the steps into position to form a complete and rigid unit. Small angle blocks can be glued at the junction of tread and riser in a step to reduce the risk of slight movement giving rise to the

Continuous handrail if total rise is over 600 mm and to both sides if stairs over 1.000 wide

900 min.
900 min.
2.000 min.
900 min.
1.000 max.

Going 220 min.

No riser 16 min.

Pitch line

Pitch line

42° max.

Rise 220 max.

Sum of going + twice rise = 550 min. to 700 max. in any flight all risers of equal height and all goings of equal width

Private stairways

Continuous handrail above 2 bottom steps and to both sides if stairs over 1.100 wide

1.100 min.
1.100 min.
2.000 min.
900 min.
1.000 max.

Going 280* (250) min.

No riser 16 min.

Pitch line

Pitch line

Rise 180 (190) max.

* 250 mm if the floor area < 100 m²

Sum of going + twice rise = 550 min. to 700 max. in any flight all risers of equal height and all goings of equal width

Institutional and assembly and (other) stairways

Figure 7.4.1   Timber stairs and Approved Document K

annoyance of creaking. The flight can be given extra rigidity by using triangular brackets placed under the steps on the centre line of the flight. The use of a central beam or carriage piece with rough brackets as a support is only used on wide stairs over 1,200 mm, especially where they are intended for used as a common stairway (see Fig. 7.4.2).

Treads 25–38 mm thick

Risers 10–12 mm thick

String position

Triangular glue blocks

Triangular brackets on centre line of stairs

String 25–38 mm thick

Wedges

15 mm deep housings

Splayed nosing

Tread

Riser

Housing joint

Rounded nosing

Tread

Riser

Rebate joint

Alternative step details

Tread

Stairs over 1.500 wide

Riser

String

Square nosing

150 × 25 rough brackets

100 × 75 carriage piece on centre line of stairs

Figure 7.4.2   Stair construction details

70 × 45 handrail

Cap

Mortice and housing for handrail

90

100

Outer string

20 × 20 balusters

100 × 100 newel post

25 mm thick string capping

100 × 100 newel post

Dowel hole

Plasterboard soffit

Mortice and housing for string

Stairs

Housing for treads and risers

Lead or similar pad

Bottom riser turned into newel

Typical detail at bottom newel

Newel notched 15 mm over trimmer

20 × 20 balusters

Skirting

T & G flooring

Wall string hooked over trimmer

Joist

90

Rebated nosing

Ceiling finish

Stairs

Trimmer

Packing

Pendant

Trimmer

Stairs

Ceiling finish

Outer string

Typical details at landing

Figure 7.4.3   Stair support and fixing details

Stairs can be designed to be either fixed to a wall with one outer string, fixed between walls or free standing – the majority have one wall string and one outer string. The wall string is fixed directly to the wall along its entire length or is fixed to timber battens plugged to the wall, the top of the string being cut and hooked over the trimming member of the stairwell. The outer string is supported at both ends by a newel post which in the case of the bottom newel rests on the floor; and, in the case of the upper newel, it is notched over and fixed to the stairwell trimming member. If the upper newel is extended to the ground floor to give extra support, it is called a storey newel post. The newel posts also serve as the termination point for handrails which span between them and is then infilled with balusters, balustrade or a solid panel to complete the protection to the sides of the stairway (see Fig. 7.4.3).

If the headroom distance is critical it is possible to construct a bulkhead arrangement over the stairs as shown in Fig. 7.4.4; this may give the increase in headroom required to comply with the Building Regulations. The raised floor in the room over the stairs can be used as a shelf or form the floor of a hanging cupboard.

LAYOUT ARRANGEMENTS

A stair flight incorporating landings or tapered steps will enable the designer to economize with the space required to accommodate the stairs. Landings can be quarter space giving a 90° turn or half space giving a 180° turn; for typical arrangements see Fig. 7.4.5. The construction of the landing is similar to that of a timber upper floor except that with the reduced span joist depths can be reduced (see Fig. 7.4.5). The landing can be incorporated in any position up the flight and if sited near the head may well provide sufficient headroom to enable a cupboard or cloakroom to be constructed below the stairs. A dog leg or string-over-string stair is economical in width since it will occupy a width less than two flights but this form has the disadvantage of a discontinuous handrail since this abuts to the underside of the return or upper flight.

TAPERED STEPS

Prior to the introduction of the Building Regulations tapered steps or winders were frequently used by designers to use space economically since three treads occupied the area required for the conventional quarter space landing which is counted as one tread. These steps had the following disadvantages:

■ Hazard to the aged and very young because of the very small tread length at or near the newel post.
■ Difficult to carpet, requiring many folds or wasteful cutting.
■ Difficult to negotiate with furniture due to a rapid rise on the turn.
■ Expensive to construct.

Typical stairwell finishes

Typical bulkhead over stairs

Figure 7.4.4   Stairwell finishes and bulkhead details

With the introduction of the Building Regulations special attention has been given to the inclusion of tapered steps in Approved Document K which makes the use of tapered steps less of an economic proposition and more difficult to design (see Fig. 7.4.6).

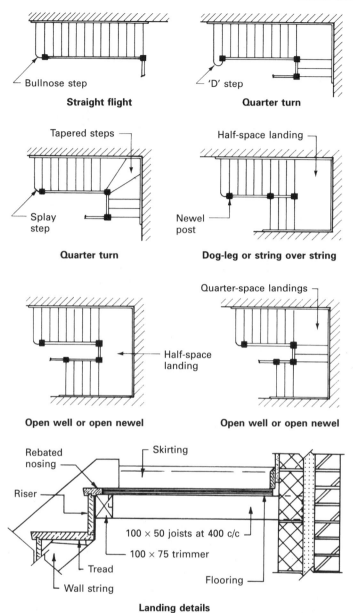

Figure 7.4.5 Typical stair layouts

OPEN TREAD STAIRS

These are a contemporary form of stairs used in homes, shops and offices based on the simple form of access stair which has been used for many years in industrial premises. The concept is simplicity with the elimination of elaborate nosings,

Max. going + twice rise = 700

Headroom clearance measured from notional width line

230

Note: Going and rise measured on centre line for stairs of less than 1.000 wide

50

Pitch lines

50

Min. going + twice rise = 550

100 × 100 newel

270

270

℄ Handrail

Width
1.000 or more

Wall string

Outer string

Tapered steps

Newel post

Easing to take bottom edge of riser

Note: Other tapered step arrangements are possible within the requirements of the Building Regulations

Wall string

Figure 7.4.6   Tapered steps for private stairways

cappings and risers. The open tread or riserless stair must fully comply with Part K of the Building Regulations and in particular to Approved Document K which recommends a minimum tread overlap of 16 mm. Where they are likely to be used by children under 5 years, i.e. dwellings, pre-school nurseries, etc., the opening between treads must contain a barrier sufficient to prevent a 100 mm sphere passing through.

Four basic types of open tread stairs can be produced:

Closed string   This will terminate at the floor and landing levels and be fixed as for traditional stairs. The treads are tightly housed into the strings which are tied together with long steel tie bars under the first, last and every fourth tread. The nuts and washers can be housed into the strings and covered with timber inserts (see Fig. 7.4.7).

Cut strings or carriages   These are used to support cantilever treads and can be worked from the solid or of laminated construction. The upper end of the carriage can be housed into the stairwell trimming member with possible additional support from metal brackets. The foot of the carriage is housed in a purpose made metal shoe or fixed with metal angle brackets (see Fig. 7.4.8).

Mono-carriage   Sometimes called a spine beam, this employs a single central carriage with double cantilever treads. The carriage, which is by necessity large, is of laminated construction and very often of a tapered section to reduce the apparent bulky appearance. The foot of the carriage is secured with a purpose made metal shoe in conjunction with timber connectors (see Fig. 7.4.9).

Alternating tread stairs   These make economic use of space and are frequently applied to loft conversions. They have a pitch of about 60° and paddle-shaped treads, which are both controversial issues in the interests of user safety. However, the Building Regulations, Approved Document K, accepts 'familiarity and regular use' as a reasonable safety argument provided the stair accesses only one habitable room. Additional requirements include a non-slip surface, handrails both sides, a minimum going of 220 mm with a maximum rise of 220 mm and a gap between treads of no more than 100 mm if likely to be used by children under 5 years (see Fig. 7.4.10).

Treads

These must be of adequate thickness since there are no risers to give extra support; usual thicknesses are 38 and 50 mm. To give a lighter appearance it is possible to taper the underside of the treads at their cantilever ends for a distance of 225–250 mm. This distance is based on the fact that the average person will ascend or descend a stairway at a distance of about 250 mm in from the handrail.

Balustrading to landing

Top newel post notched over trimmer

Plank balustrade fixed to face of newel

Skirting to landing

Wall string hooked over trimmer

Treads housed into strings

Apron lining

Stairwell trimmer

Outer string

Bottom newel post

Typical arrangement

Exposed bolt head

Tread

String

12 mm deep housing

Tie bar every 4th tread

Wood cover insert often diamond shape for neat appearance

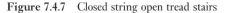

Alternative tie bar arrangements

Figure 7.4.7 Closed string open tread stairs

Typical arrangement

Labels in figure:
- Stepped rebated nosing
- Wrot-faced trimmer
- Gluelam timber handrail
- Handrail to landing
- Tread
- Cut string or carriage
- Timber knee board
- Metal standards or balusters
- Fixing bracket
- Treads cantilevered 225 mm over carriage at ends
- Stepped rebated nosing
- Wrot-faced trimmer
- Carriage housed into trimmer

Figure 7.4.8 Cut string open tread stairs

Handrail to landing

Stepped
rebated nosing

Wrot-faced
trimmer

Timber or
metal
handrail

Tread

Laminated timber
mono-carriage
housed into trimmer

Metal balustrade

Typical arrangement

Fixing
shoe

Laminated
timber
carriage

Purpose-
made cast
iron or steel
shoe

Holes for
through bolt
using single
toothed timber
connectors
between shoe
and carriage

Holes for
holding-down bolts

Figure 7.4.9   Mono–carriage open tread stairs

Plan

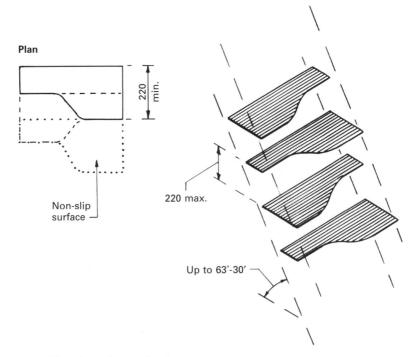

220 min.

Non-slip
surface

220 max.

Up to 63°-30′

Figure 7.4.10 Alternating tread stairs

Balustrading

Together with the handrail balustrading provides both the visual and practical
safety barrier to the side of the stair. Children present special design problems since
they can and will explore any gap big enough to crawl through. BS 5395 for wood
stairs recommends that the infill under handrails should have no openings which
would permit the passage of a sphere 90 mm in diameter, a slight reduction on
the 100 mm stated in Approved Document K1, Section 1.29, to the Building
Regulations. Many variations of balustrading are possible ranging from simple
newels with planks to elaborate metalwork of open design (see Figs 7.4.7, 7.4.8,
and 7.4.9).

SIMPLE REINFORCED CONCRETE STAIRS

The functions of any stairway are:

1. To provide for the movement of people from one floor level to another.
2. To provide a degree of sound insulation where part of a separating element between compartments in a building. (Building Regulations, Parts E2 and E3.)
3. To provide a suitable means of escape in event of fire. (Building Regulations, Part B1.)

A timber stairway will maintain its strength and integrity for a reasonable period during an outbreak of fire. However, timber is combustible and will contribute to the fire spread, thus increasing the hazards which could be encountered along an escape route. It is for this reason that application of timber stairs are limited by the Building Regulations to certain domestic and some other small building types. Stairs in this category are not normally in a protected shaft and therefore not part of a specifically planned escape route.

Stairs, other than the exceptions given above, must therefore be constructed of non-combustible materials, but combustible materials are allowed to be used as finishes to the upper surface of the stairway or landing. Reinforced concrete stairs are non-combustible, strong and hard wearing. They may be constructed *in-situ* or precast in sections ready for immediate installation and use when delivered to site. The general use of cranes on building sites has meant that many of the large flight arrangements, which in the past would have been cast *in-situ*, can now be precast under factory-controlled conditions.

Many variations of plan layout and spanning direction are possible but this study will be confined to the simple straight flight spanning from floor to floor or floor to intermediate landing. The designer will treat the stairs as being an inclined slab spanning simply between the supports, the steps being treated as triangular loadings evenly distributed over the length. Where intermediate landings are included in the design the basic plan is similar to the open well or newel timber stairs. Difficulty is sometimes experienced with the intersection of the upper and lower flight soffits with the landing. One method of overcoming this problem is to have, in plan, the top and bottom steps out of line so that the soffit intersections form a straight line on the underside of the landing (see Fig. 7.5.1). The calculations to determine the

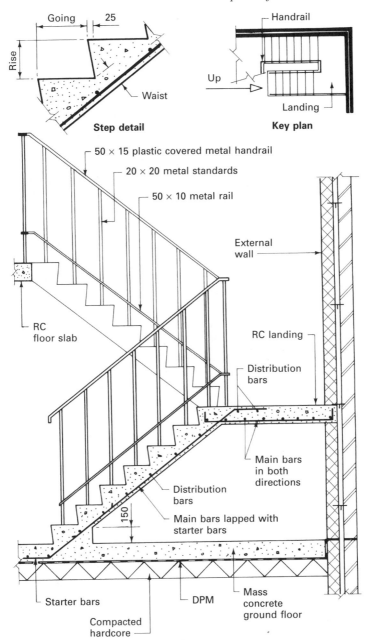

Figure 7.5.1 Simple reinforced concrete *in-situ* stairs

rise, going and number of steps are the same as those used for timber stairs; it should be noted that the maximum number of risers in each flight must not exceed sixteen. To achieve a greater tread length without increasing the actual horizontal going it is common to use a splayed riser face giving a 25 mm increase to the tread length.

The concrete specification should be very strong, with a mix ratio of 1:1.5:3–10 mm aggregate or 25–30 N/mm^2 compressive strength. Water/cement ratio is no greater than 0.5, otherwise concrete will flow over the formwork riser boards. Concrete cover to reinforcement is at least 15 mm or the bar diameter, whichever is the greater. This cover is sufficient for one-hour fire resistance in most buildings, but will need to be increased for buildings in certain purpose groups – see Table A2 in Appendix A of Approved Document B to the Building Regulations. The thickness of concrete required is dependent on the loading and span but is not generally less than 100 mm or more than 150 mm measured across the waist which is the distance from the soffit to the intersection of tread and riser (see Fig. 7.5.1).

Mild steel or high yield steel bars can be used to reinforce concrete stairs, the bars being lapped to starter bars at the ground floor and taken into the landing or floor support slab. The number, diameter and spacing of the main and distribution reinforcement must always be calculated for each stairway.

Handrails and balustrading must be constructed of a non-combustible material, continuous and to both sides if the width of the stairs exceeds 1.000 m. The overall height of the handrail up the stairs should be between 900 and 1,000 mm measured vertically and have a height above the floor of 1.100 m minimum. The capping can be of a combustible material such as plastic provided that it is fixed to or over a non-combustible core. Methods of securing balustrades and typical handrail details are shown in Fig. 7.5.2. Flights of over 1.8 m width in public buildings should have a centrally dividing handrail. Widths of stairs designated for fire escape purposes should be determined by reference to tables and calculations in Section 4 of the Building Regulations, Approved Document B1.

A wide variety of finishes can be applied to the tread surface of the stairs. If the appearance is not of paramount importance, such as in a warehouse, a natural finish could be used but it would be advisable to trowel into the surface some carborundum dust to provide a hard-wearing non-slip surface. Alternatively, rubber or carborundum insert strips could be fixed or cast into the leading edges of the treads. Finishes such as PVC tiles, rubber tiles, and carpet mats are applied and fixed in the same manner as for floors. The soffits can be left as struck from the formwork and decorated or finished with a coat of spray plaster or a skim coat of finishing plaster.

FORMWORK

The basic requirements are the same as for formwork to a framed structure. The stair profile is built off an adequately supported soffit of sheet material by using a cut string. Riser boards are used to form the leading face of the steps; these should have a splayed bottom edge to enable complete trowelling of the tread surfaces and to ensure that air is not trapped under the bottom edge of the riser board thus

Typical handrails

Typical fixing methods

Figure 7.5.2 Handrails and balustrades

causing voids. If the stair abuts a vertical surface two methods can be considered to provide the abutment support for the riser boards; a reversed cut string or a wall board with hangers (see Fig. 7.5.3). Wide stairs can have a reverse cut string as a central support to the riser boards to keep the thickness of these within an acceptable coat limit.

200 × 38 wall board fixed to wall
75 × 50 hanger
50 × 32 struts
21 mm plywood soffit board
21 mm plywood risers
150 × 50 joists
75 × 32 ribbon piece
15 mm plywood cut string
Alternative wall support – reversed cut string

21 mm plywood soffit to landing
Wall board
Hangers
Riser boards
Joists supported by adjustable steel props
2 No. 150 × 75 carriage pieces
Joists
Sole plate
Cut string

Figure 7.5.3 Typical formwork to reinforced concrete *in-situ* stairs

SIMPLE PRECAST CONCRETE STAIRS

Precast concrete stairs can be designed and constructed to satisfy a number of different requirements. They can be a simple inclined slab, a cranked slab, open riser stairs or constructed from a series of precast steps built into, and if required cantilevered from, a structural wall.

The design considerations for the simple straight flight are the same as those for *in-situ* stairs of comparable span, width and loading conditions. The fixing and support, however, require a different approach. Bearings for the ends of the flights must be provided at the floor or landing levels in the form of a haunch, rebate or bracket and continuity of reinforcement can be achieved by leaving projecting bars and slots in the floor into which they can be grouted (see Fig. 7.6.1).

Ideally the delivery of precast stairs should be arranged so that they can be lifted, positioned and fixed direct from the delivery vehicle, thus avoiding double handling. Precast components are usually designed for two conditions:

1. Lifting and transporting.
2. Final fixed condition.

It is essential that the flights are lifted from the correct lifting points, which may be in the form of loops or hooks projecting from or recessed into the concrete member, if damage by introducing unacceptable stresses during lifting is to be avoided.

Balustrade and handrail requirements together with the various methods of fixing are as described for *in-situ* reinforced concrete stairs. Any tread finish which is acceptable for an *in-situ* stair will also be suitable for the precast alternative.

The use of precast concrete steps to form a stairway is limited to situations such as short flights between changes in floor level and external stairs to basements and areas. They rely on the loadbearing wall for support and if cantilevered on the downward load of the wall to provide the necessary reaction. The support wall has to provide this necessary load and strength, and at the same time it has to be bonded or cut around the stooled end(s) of the steps. It is for these reasons that the application of precast concrete steps is restricted.

319

Precast concrete flight of stairs

Projecting bonding
or continuity bars

Projecting
bonding or
continuity bars

Chases to receive
bonding bars

Rebate for precast
concrete flight

Chases in floor
made good with
matching grout

RC
floor
slab

RC edge
beam

Flight in position

RC landing

RC edge beam

Simple precast concrete stairs

Square or
stooled end
built into
wall

Cantilever step

Support wall

Precast concrete
cantilever steps
built into wall
forming stairs

Simple precast concrete steps

Figure 7.6.1   Precast concrete stairs and steps

PARTITIONS

Internal walls which divide the interior of a building into areas of accommodation and circulation are called *partitions* and these can be classified as loadbearing or non-loadbearing partitions.

LOADBEARING PARTITIONS

These are designed and constructed to receive superimposed loadings and transmit these loads to a foundation (see Fig. 4.1.2). Generally loadbearing partitions are constructed of bricks or blocks bonded to the external walls (see Figs 3.1.20 and 3.1.21). Openings are formed in the same manner as for external walls; a lintel spans the opening carrying the load above to the reveals on either side of the opening. Fixings can usually be made direct into block walls, whereas walls of bricks require to be drilled and plugged to receive nails or screws. Apart from receiving direct fixings block partitions are lighter, cheaper, quicker to build and have better thermal insulation properties than brick walls but their sound insulation values are slightly lower. For these reasons blockwork is usually specified for loadbearing partitions in domestic work. Loadbearing partitions, because of their method of construction, are considered to be permanently positioned.

NON-LOADBEARING PARTITIONS

These partitions, like loadbearing partitions, must be designed and constructed to carry their own weight and any fittings or fixings which may be attached to them, but they must not, in any circumstances, be used to carry or assist in the transmission of structural loadings. They must be designed to provide a suitable division between internal spaces and be able to resist impact loadings on their faces, also any vibrations set up by doors being closed or slammed.

Bricks or blocks can be used for the construction of non-loadbearing partitions. They may be built directly off a concrete floor with a thickened slab or steel reinforcement to support the wall's self-weight. At ceiling level stability is obtained by bracketing between or to the joists above. If the partition is built off a suspended

Head plate   Head to opening

Noggins

Studs   Jambs to opening   Folding wedges

Sole plate

Typical arrangement

95 × 70 jamb to opening   12 mm deep housing

90 × 45 head plate fixed to joists

Tenon
Wedge

Mortice

95 × 70 head to opening

Rebated door lining

Architrave

Skirting

Lining

95 × 45 studs at 400 c/c

90 × 45 sole plate bolted to floor

Figure 7.7.1   Typical timber stud partition

Table 7.7.1 Guidance when using galvanized steel taper nails

Plasterboard thickness	Plasterboard width	Stud or batten spacing	Nails
9.5	900	450	2 × 30 long
	1,200	400	2 × 30 long
12.5	600	600	2 × 40 long
	900	450	2 × 40 long
19.0	1,200	600	2.6 × 40 long
	600	600	2.6 × 40 long

Note: All dimensions in millimetres.

timber floor a larger joist section or two joists side by side must be used to carry the load of the partition to the floor support walls. Openings are constructed as for loadbearing walls or alternatively a storey height frame could be used.

Timber stud partitions with suitable facings are lighter than brick or block partitions but are less efficient as sound or thermal insulators. However, a high standard of sound and thermal insulation and fire resistance can be achieved by facing with plasterboard, filling the voids between studs and noggins with glass or mineral wool. They are easy to construct and provide a good fixing background and because of their lightness are suitable for building off a suspended timber floor. The basic principle is to construct a simple framed grid of timber to which a dry lining such as plywood, plasterboard or hardboard can be attached. The lining material will determine the spacing of the uprights or studs to save undue wastage in cutting the boards to terminate on the centre line of a stud. To achieve a good finish it is advisable to use studs which have equal thicknesses since the thin lining materials will follow any irregularity of the face of the studwork. Openings are formed by framing a head between two studs and fixing a lining or door frame into the opening in the stud partition; typical details are shown in Fig. 7.7.1. Where plasterboard is the specified facing/lining material, the extent of nailing will depend on the stud spacing, the facing thickness and its width. Table 7.7.1 provides guidance when using galvanized steel taper head nails at 150 mm spacing.

Proprietary partitions of plasterboard bonded on either side of a strong cellular core to form rigid panels are suitable as non-loadbearing partitions. These are fixed to wall and ceiling battens and supported on a timber sole plate. Timber blocks are inserted into the core to provide the fixing medium for door frames or linings and skirtings (see Fig. 7.7.2). This form of partition is available with facings suitable for direct decoration, plaster coat finish or with a plastic face as a self-finish.

Compressed strawboard panels provide another alternative method for non-loadbearing partitions. The storey height panels are secured to a sole plate and a head plate by skew or tosh nailing through the leading edge. The 3 mm joint between consecutive panels is made with an adhesive supplied by the manufacturer. Openings are formed by using storey height frames fixed with 100 mm screws direct into the edge of the panel. The joints are covered with a strip of hessian scrim and the whole partition is given a skim coat of board plaster finish (see Fig. 7.7.2).

50 mm thick 'Paramount' cellular
plasterboard partition panel

30 × 20 head batten
fixed to ceiling

Spandrel
panel over
opening

30 × 20 batten
fixed
to wall

30 × 20 spandrel
fixing batten

30 × 20 door frame
fixing batten

Door frame

Planted door
stop

50 × 20 sole
plate fixed
to floor

Skirting

Holes at 300 c/c for electric wiring

3 mm adhesive joint

75 mm wide head
plate fixed
between
joists

Hessian
scrim

50 mm thick
'stramit'
compressed
strawboard
panels

Plaster
skim coat

Storey height
frame

Threshold

Sole plate
fixed to floor
and grooved
for electric
wiring

Skirting

Architrave

Figure 7.7.2   Typical preformed partitions

Partially prefabricated partitions, such as the plasterboard and strawboard described above, can be erected on site without undue mess. Being mainly a dry construction they reduce the drying time required with the traditional brick and block walls. They also have the advantage, like timber stud partitions, of being capable of removal or repositioning without causing serious damage to the structure or without causing serious problems to a contractor.

FINISHES: FLOOR, WALL AND CEILING

FLOOR FINISHES

The type of floor finish to be applied to a floor will depend upon a number of factors such as type of base, room usage, degree of comfort required, maintenance problems, cost, appearance, safety and individual preference.

Floor finishes can be considered under three main headings:

1. *In-situ* floor finishes – those finishes which are mixed on site, laid in a fluid state, allowed to dry and set to form a hard jointless surface.
2. Applied floor finishes – those finishes which are supplied in tile or sheet form and are laid on to a suitably prepared base.
3. Timber floor finishes – boards, sheets and blocks of timber laid on or attached to a suitable structural frame or base.

IN-SITU FLOOR FINISHES

Mastic asphalt – BS 6925 and BS 8204: Part 5

This is a naturally occurring bituminous material obtained from asphalt lakes like those in Trinidad or it can also be derived from crude oil residues. Trinidad lake asphalt is used as a matrix or cement to a suitably graded mineral aggregate to form mastic asphalt as a material suitable for floor finishing. When laid mastic asphalt is impervious to water and is ideal for situations such as sculleries, wash rooms and balconies. It also forms a very good surface on which to apply thin tile and sheet finishes (for example, PVC) and will at the same time fulfil the function of a damp-proof membrane or waterproof decking.

Mastic asphalt is a thermoplastic material and has to be melted before it can be applied to form a jointless floor finish. Hot mastic asphalt is applied by means of a float at a temperature of between 180 and 210 °C in a single 13 mm coat as a base for applied finishes or in a 16 mm single coat for a self finish. Any sound and rigid subfloor is suitable but a layer of ordinary black sheathing felt should be included between the subfloor and mastic asphalt to overcome the problems caused by

326

differential movement. The finish obtained is smooth and hard but the colour range is limited to dark colours such as red, brown and black. A matt surface can be produced by giving the top surface a dusting of sand or powdered stone.

Pitch mastic

Pitch mastic is a similar material to mastic asphalt but is produced from a mixture of calcareous and/or siliceous aggregates bonded with coal tar pitch. It is laid to a similar thickness and in a similar manner to mastic asphalt with a polished or matt finish. Pitch mastic floors are resistant to water but have a better resistance to oil and fats than mastic asphalt and are therefore suitable for sculleries, washrooms and kitchens.

Granolithic

This is a mixture of Portland cement and granite chippings which can be applied to a 'green' concrete subfloor or to a cured concrete subfloor. Green concrete is a term used to describe newly laid concrete which is not more than three hours old. A typical mix for granolithic is 1 part cement: 1 part sand: 2 parts granite chippings (5–10 mm free from dust) by volume. The finish obtained is very hard wearing, noisy and cold to touch; it is used mainly in situations where easy maintenance and durability are paramount such as a common entrance hall to a block of flats.

If granolithic is being applied to a green concrete subfloor as a topping it is applied in a single layer approximately 20 mm thick in bay sizes not exceeding 28 m^2 and trowelled to a smooth surface. This method will result in a monolithic floor and finish construction.

The surface of mature concrete will need to be prepared by hacking the entire area and brushing well to remove all the laitance before laying a single layer of granolithic which should be at least 40 mm thick. The finish should be laid, on a wet cement slurry coating or PVA bonding agent to improve the bond, in bay sizes not exceeding 14 m^2.

Magnesium oxychloride

This is a composition flooring which is sometimes used as a substitute for asphalt since it has similar wearing and appearance properties. It is mixed on site using a solution of magnesium chloride with burned magnesite and fillers such as wood flour, sawdust, powdered chalk or limestone. It is essential that the ingredients are thoroughly mixed in a dry state before the correct amount of solution is added. The mixed material is laid in one or two layers giving a total thickness of approximately 20 mm. The subfloor must be absolutely dry otherwise the finish will lift and crack; therefore a damp-proof membrane in the subfloor is essential.

APPLIED FLOOR FINISHES

Many of the applied floor finishes are thin flexible materials and should be laid on a subfloor with a smooth finish. This is often achieved by laying a cement/sand bed or screed with a steel float finish to the concrete subfloor. The usual screed mix is 1:3 cement/sand and great care must be taken to ensure that a good bond is obtained with the subfloor. If the screed is laid on green concrete a thickness of 12 mm is usually specified whereas screeds laid on mature concrete require a thickness of 40 mm. A mature concrete subfloor must be clean, free from dust and dampened with water to reduce the suction before applying the bonding agent to receive the screed. To reduce the possibility of drying shrinkage cracks screeds should not be laid in bays exceeding 15 m² in area. Screeds laid directly over rigid insulation should be at least 50 mm in thickness, reinforced with light steel mesh (chicken wire).

Flexible PVC tiles and sheet – BS 3261 and BS 8203

Flexible PVC is a popular hardwearing floor finish produced by a mixture of polyvinyl chloride resin, pigments, mineral fillers and plasticizers to control flexibility. It is produced as 300×300 mm square tiles or in sheet form up to 2,400 mm wide with a range of thicknesses from 1.5 to 4.5 mm. The floor tiles and sheet are fixed with an adhesive recommended by the manufacturer and produce a surface suitable for most situations. PVC tiles, like all other small unit coverings, should be laid from the centre of the area towards the edges, so that if the area is not an exact tile module an even border of cut tiles is obtained.

Thermoplastic tiles

These are sometimes called asphalt tiles and are produced from coumarone indene resins, fillers and pigments; the earlier use of asphalt binders limited the range of colours available. These tiles are hardwearing, moisture resistant and are suitable for most situations, being produced generally as a 225 mm square tile either 3 or 4.5 mm thick. To make them pliable they are usually heated before being fixed with a bituminous adhesive. As they age they can become brittle and fracture, detaching from their base adhesive. Development of PVC tiles has largely displaced thermoplastic tiles as they are more resilient and offer a brighter colour range.

Rubber tiles and sheet – BS 1711 and BS 3187

Solid rubber tiles or sheet is produced from natural or synthetic rubber compounded with fillers to give colour and texture, the rubber content being not less than 35%. The covering is hardwearing, quiet and water resistant and suitable for bathrooms and washrooms. Thicknesses range from 3 to 6.5 mm with square tile sizes ranging from 150 to 1,200 mm; sheet widths range from 900 to 1,800 mm. Fixing is by rubber-, epoxy- or neoprene-based adhesives, as recommended by the manufacturer, to a smooth surface.

Linoleum – BS 6826

Linoleum is produced in sheet or tile form from a mixture of drying oils, resins, fillers and pigments which is pressed on to a hessian or bitumen saturated felt paper backing. Good-quality linoleum has the pattern inlaid or continuing through the thickness whereas the cheaper quality has only a printed surface pattern. Linoleum gives a quiet, resilient and hardwearing surface suitable for most domestic floors. Thicknesses vary from 2 to 6.5 mm for a standard sheet width of 1,800 mm; tiles are usually 300 mm square with the same range of thicknesses. Fixing of linoleum tiles and sheet is by adhesive to any dry smooth surface, although the adhesive is sometimes omitted with the thicker sheets.

Carpet – BS 4223, BS 5808 and BS 5325

The chief materials used in the production of carpets are nylon, acrylics and wool or mixtures of these materials. There is a vast range of styles, types, patterns, colours, qualities and sizes available for general domestic use in dry situations since the resistance of carpets to dampness is generally poor. To obtain maximum life carpets should be laid over an underlay of felt or latex and secured by adhesives, nailing around the perimeter or being stretched over and attached to special edge fixing strips. Carpet is supplied in narrow or wide rolls, carpet squares (600 × 600 × 25 mm thick) are also available for covering floors without the use of adhesives; these rely on the interlocking of the edge fibres to form a continuous covering.

Cork tiles and carpet – BS 6826

Cork tiles are cut from baked blocks of granulated cork where the natural resins act as the binder. The tiles are generally 300 mm square with thicknesses of 5 mm upwards according to the wearing quality required and are supplied in three natural shades. They are hardwearing, quiet and resilient but unless treated with a surface sealant they may collect dirt and grit. Fixing is by using an adhesive recommended by the manufacturer, with the addition of nine steel pins around the edge to prevent curling.

Cork carpet is a similar material but it is made pliable by bonding the cork granules with linseed oil and resins on to a jute canvas backing. It is laid in the same manner as described above for linoleum and should be treated with a surface sealant to resist dirt and grit penetration.

Quarry tiles – BS 6431

The term *quarry* is derived from the Italian word *quadro* meaning square and does not mean that the tiles are cut or won from an open excavation or quarry. They are made from ordinary or unrefined clays worked into a plastic form, pressed into shape and hard burnt. Being hardwearing and with a good resistance to water they are suitable for kitchens and entrance halls but they tend to be noisy and cold. Quarry tiles are produced as square tiles in sizes ranging from $100 \times 100 \times 20$ mm to $225 \times 225 \times 32$ mm thick.

Three methods of laying quarry tiles are recommended to allow for differential movement due to drying shrinkage or thermal movement of the screed. The first method is to bed the tiles in a 10–15 mm thick bed of cement mortar over a separating layer of sheet material such as polythene, which could perform the dual function of damp-proof membrane and separating layer. To avoid the use of a separating layer the 'thick bed' method can be used. With this method the concrete subfloor should be dampened to reduce suction and then covered with a semi-dry 1:4 cement/sand mix to a thickness of approximately 40 mm. The top surface of the compacted semi-dry 'thick bed' should be treated with a 1:1 cement/sand grout before tapping the tiles into the grout with a wood beater. Alternatively a dry cement may be trowelled into the 'thick bed' surface before bedding in the tiles. Special cement-based adhesives are also available for bedding tiles to a screed and these should be used in accordance with the manufacturer's instructions to provide a thin bed fixing.

Clay tiles may expand probably due to physical adsorption of water and chemical hydration and for this reason an expansion joint of compressible material should be incorporated around the perimeter of the floor (see Fig. 7.8.1). In no circumstances should the length or width of a quarry tile floor exceed 7,500 mm without an expansion joint. The joints between quarry tiles are usually grouted with a 1:1 cement/sand grout and then, after cleaning, the floor should be protected with sand or sawdust for four or five days. There is a wide variety of tread patterns available to provide a non-slip surface, also a wide range of fittings are available to form skirtings and edge coves (see Fig. 7.8.1).

Plain clay or ceramic floor tiles – BS 6431

These are similar to quarry tiles but are produced from refined natural clays which are pressed after grinding and tempering into the desired shape before being fired at a high temperature. Plain clay floor tiles, being denser than quarry tiles, are made as smaller and thinner units ranging from 50×50 mm to 300×300 mm in thicknesses of 9.5 to 13 mm.

Laying, finishes and fittings available are all as described for quarry tiles.

TIMBER FLOOR FINISHES

Timber is a very popular floor finish with both designer and user because of its natural appearance, resilience and warmth. It is available as a board, strip, sheet or block finish and if attached to joists, as in the case of a suspended timber floor, it also acts as the structural decking.

Studded pattern anti-slip tile

150 mm high quarry tile skirting with rounded top edge and coved bottom edge

150 × 150 quarry tiles

Grouted joints (1:2)

Separating layer

Partition wall

Wall finish

Cement/sand bed (1:3)

Expansion joint

Four-pannelled anti-slip tile

150 mm high quarry tile skirting with rounded top edge and square bottom edge

150 × 150 quarry tiles

Wall finish

Grouted joints (1:2)

Expansion joint

40 mm thick cement/sand bed (1:4)

Figure 7.8.1   Typical quarry tile floors

Timber boards

Softwood timber floor boards are joined together by tongued and grooved joints along their edges and are fixed by nailing to the support joists or fillets attached to a solid floor. The boards are butt jointed in their length, the joints being positioned centrally over the supports and staggered so that butt joints do not occur in the same position in consecutive lengths. The support spacing will be governed by the spanning properties of the board (which is controlled by its thickness) and supports placed at 400 mm centres are usual for boards of 19 and 22 mm thick. The tongue is positioned slightly off centre and the boards should be laid with the narrow shoulder on the underside to give maximum wear. It is essential that the boards are well cramped together before being fixed to form a tight joint and that they are laid in a position where they will not be affected by dampness. Timber is a hygroscopic material and will therefore swell and shrink as its moisture content varies – ideally this should be maintained at around 12%.

Timber strip

These are narrow boards being under 100 mm wide to reduce the amount of shrinkage and consequent opening of the joints. Timber strip can be supplied in softwood or hardwood and is considered to be a superior floor finish to boards. Jointing and laying is as described for boards, except that hardwood strip is very often laid one strip at a time and secret nailed (see Fig. 7.8.2).

Timber sheet floor finish – BS 5669

Chipboard is manufactured from wood chips or shavings bonded together with thermosetting synthetic resins, and forms rigid sheets 18 and 22 mm thick which is suitable as a floor finish. The sheets are fixed by nailing or screwing to support joists or fillets. If the sheet is used as an exposed finish a coating of sealer should be used. Alternatively chipboard can be used as a decking to which a thin tile or carpet finish can be applied.

Tongued and grooved boards of 600 mm width are also available as a floor-decking material which does not need to be jointed over a joist.

Wood blocks

These are small blocks of timber usually of hardwood which are designed to be laid in set patterns. Lengths range from 150 to 300 mm with widths up to 89 mm; the width being proportional to the length to enable the various patterns to be created. Block thicknesses range from 20 to 30 mm thick and the final thickness after sanding and polishing being about 5 to 10 mm thinner.

The blocks are jointed along their edges with a tongued and grooved joint and have a rebate, chamfer or dovetail along the bottom longitudinal edges to take up any surplus adhesive used for fixing. Two methods can be used for fixing wood

90

20

Secret nailing

Typical strip section

Wall finish

Partition wall

Skirting

Hardwood strip floor
secret nailed to
timber fillets

Solid concrete
floor to comply
with B. Reg. C4

3 mm thick sandwich
membrane of bitumen
not lower than highest
adjoining ground level

Pressure
impregnated
timber fillets
at 450 c/c

Wall finish

Tongue

Rebate

Wall

Typical block

Herringbone pattern

Skirting

20 mm thick blocks
laid in basket
pattern

Cork
strip

Bitumen
adhesive

Solid concrete floor to comply
with B. Reg. C4 with floated finish

Figure 7.8.2 Typical timber floor finishes

blocks: the first uses hot bitumen and the second a cold latex bitumen emulsion. If hot bitumen is used the upper surface of the subfloor is first primed with black varnish to improve adhesion, and then, before laying, the bottom face of the block is dipped into the hot bitumen. The cold adhesive does not require a priming coat to the subfloor. Blocks, like tiled floors, should be laid from the centre of the floor towards the perimeter, which is generally terminated with a margin border. To allow for moisture movement a cork expansion strip should be placed around the entire edge of the block floor (see Fig. 7.8.2).

Parquet

This is a superior form of wood block flooring made from specially selected hardwoods chosen mainly for their decorative appearance. Parquet blocks are generally smaller and thinner than hardwood blocks and are usually fixed to a timber subfloor which is level and smooth. Fixing can be by adhesives or secret nailed, alternatively they can be supplied as a patterned panel fixed to a suitable backing sheet in panel sizes from 300 to 600 mm square.

WALL FINISHES

External brickwork with an exposed face of facing bricks is a self-finish and requires no further treatment. External walls of common bricks or blocks can be treated to give an acceptable appearance by the application of paint, an applied wall finish such as rendering or be clad with boards or tiles. Internal walls or partitions can be built with a fair face of natural materials such as bricks or stone but generally it is cheaper to use a material such as blocks with an applied finish like plaster, dry lining or glazed tiles.

EXTERNAL RENDERING – BS 5262

This is a form of plastering using a mixture of cement and sand, or cement, lime and sand, applied to the face of a building to give extra protection against the penetration of moisture or to provide a desired texture. It can also be used in the dual capacity of providing protection and appearance.

The rendering must have the properties of durability, moisture resistance and an acceptable appearance. The factors to be taken into account in achieving the above requirements are mix design, bond to the backing material, texture of surface, degree of exposure of the building and the standard of workmanship in applying the rendering.

Cement and sand mixes will produce a strong moisture-resistant rendering but one which is subject to cracking due to high drying shrinkage. These mixes are used mainly on members which may be vulnerable to impact damage such as columns. Cement, lime and sand mixes have a lower drying shrinkage but are more absorbent than cement and sand mixes; they will, however, dry out rapidly after periods of rain and are therefore the mix recommended for general use.

The two factors which govern the proportions to be used in a mix are:

- Background to which the rendering is to be applied.
- Degree of exposure of the building.

The two common volume mix ratios are:

- $1:\frac{1}{2}:4-4\frac{1}{2}$ cement:lime:sand which is used for dense, strong backgrounds of moderate to severe exposure and for application to metal lathing or expanded metal backgrounds. Equivalent cement and sand with a plasticizer is 1:3–4.
- 1:1:5–6 cement:lime:sand which is for general use. Equivalent cement and sand with a plasticizer is 1:5–6.

If the rendering is to be applied in a cold, damp situation the cement content of the mix ratio should be increased by 1. The final coat is usually a weaker mix than the undercoats, for example, when using an undercoat mix 1:1:6 (1:5–6 cement: sand with plasticizer), the final coat mix ratio should be 1:2:9 (1:7–8 cement:sand with plasticizer), to reduce the amount of shrinkage in the final coat. The number of coats required will depend upon the surface condition of the background and the degree of exposure. Generally a two-coat application is acceptable, except where the background is very irregular or the building is in a position of severe exposure when a three-coat application would be specified. The thickness of any one coat should not exceed 15 mm and each subsequent coat thickness is reduced by approximately 3 mm to give a final coat thickness of 6–10 mm. Wood float treatment produces a smooth flat surface which may be treated with a felt-coated float or lamb's wool roller to raise the surface texture and provide better resistance to cracking.

Various textured surfaces can be obtained on renderings by surface treatments such as scraping the surface with combs, saw blades or similar tools to remove a surface skin of mortar. These operations are carried out some three to four hours after the initial application of the rendering and before the final set takes place.

Alternative treatments are:

Roughcast A wet plaster mix of 1 part cement: $\frac{1}{2}$ part lime: $1\frac{1}{2}$ parts shingle:3 parts sand which is thrown on to a porous coat of rendering to give an even distribution.

Pebble or dry dash Selected aggregate such as pea shingle is dashed or thrown on to a rendering background before it has set and is tamped into the surface with a wood float to obtain a good bond.

Spattered finishes These are finishes applied by a machine (which can be hand operated), guns or sprays using special mixes prepared by the machine manufacturers.

PLASTERING

Plastering, like brickwork, is one of the old-established crafts in this country having been introduced by the Romans. The plaster used was a lime plaster which generally has been superseded by gypsum plasters. The disadvantages of lime plastering are:

- Drying shrinkage which causes cracking.
- Slow drying-out process which can take several weeks causing delays for the following trades.
- Need to apply lime plaster in several coats, usually termed render, float and set, to reduce the amount of shrinkage.

Any plaster finish must fulfil the functions of smoothing out irregularities in the backing wall, provide a continuous surface which is suitable for direct decoration and be sufficiently hard to resist damage by impact upon its surface; gypsum plasters fulfil these requirements.

Table 7.8.1   Categories of plaster

Gypsum-based plasters	BS ref.	Trade name
Class A:		
Plaster of Paris hemi-hydrate	BS 1191: Part 1	
Class B:		
Retarded hemi-hydrate	BS 1191: Part 1	
Undercoats:		
Browning	– section a1	Thistle browning
		Thistle slow set
		Thistle fibred
Metal lathing	– section a2	Thistle metal lathing
Finishes:		
Finish	– section b1	Thistle finish, Sirapite B
Board finish	– section b2	Thistle board finish
Class C:		
Anhydrous finish only	BS 1191: Part 1	Sirapite
Class D:		
Keene's plaster	BS 1191: Part 1	Standard Keene's
		Fine, polar
		white cement
Premixed lightweight	BS 1191: Part 2	
Undercoats:		
Browning	– section a1	Carlite browning
		Carlite browning (HSB)
Metal lathing	– section a2	Carlite metal lathing
Bonding	– section a3	Carlite bonding
		Carlite welter-weight bonding
Finishes:		
Finish	– section b1	Carlite finish, Limelight finishing

Gypsum is a crystalline combination of calcium sulphate and water. Deposits of suitable raw material are found in several parts of England and after crushing and screening the gypsum is heated to dehydrate the material. The amount of water remaining at the end of this process defines its class under BS 1191: *Specification of gypsum building plasters*. If powdered gypsum is heated to about 170 °C it loses about three-quarters of its combined water and is called hemi-hydrate gypsum plaster but is probably better known as 'Plaster of Paris'. If a retarder is added to the hemi-hydrate plaster a new class of finishing plaster is formed to which the addition of expanded perlite and other additives will form a one-coat or universal plaster, a renovating grade plaster or a spray plaster for application by a spray machine.

BS 1191 establishes four classes of plaster:

- Class A – Plaster of Paris.
- Class B – Retarded hemi-hydrate gypsum plaster in two categories:
  (a) Undercoat plasters.
  (b) Finish coat plasters.
- Class C – Anhydrous gypsum plaster (completely dehydrated gypsum).
- Class D – Keene's plaster (completely dehydrated gypsum, very slow set, high-quality finish).

Table 7.8.1 lists British Standard plaster classifications, but omits non-standard plasters for special applications. These include:

- Thin-wall plasters – for skimming and filling. These may contain organic binders which could be incompatible with some backgrounds. Check with manufacturer before application.
- Projection plasters – for spray machine application.
- X-ray plasters – contain barium sulphate aggregate. This is a heavy metal with salts that provide a degree of exposure protection.
- Acoustic plaster – contain porous gaps which will absorb sound. Surface treatments that obscure should be avoided.
- Resinous plaster – 'Proderite' formula S, a cement-based plaster especially formulated for squash courts as a substitute for Class D – Keene's plaster.

Gypsum plasters are not suitable for use in temperatures exceeding 43 °C and should not be applied to frozen backgrounds. However, plasters can be applied under frosty conditions providing the surfaces are adequately protected from freezing after application.

Gypsum plasters are supplied in multi-walled paper sacks, which require careful handling. Damage to the bags will admit moisture and commencement of the setting process. Plaster in an advanced state of set will have a short working time and may lack considerable strength. Characteristics are manifested in lack of bond and surface irregularities. Plaster should be stored in a dry location on pallets or other means of avoiding the ground. The dry, bagged plaster in store will be unaffected by low temperatures. Deterioration with age must also be avoided by regular checks on the manufacturer's date stamps and applying a system of strict rotation.

The plaster should be mixed in a clean plastic or rubber bucket using clean water only. Cleanliness is imperative since any set plaster left in the mixing bucket from a previous mixing will shorten the setting time which may reduce the strength of the plaster when set.

Premixed plasters incorporate lightweight aggregates such as expanded perlite and exfoliated vermiculite. Perlite is a glassy volcanic rock combining in its chemical composition a small percentage of water. When subjected to a high temperature the water turns into steam thus expanding the natural perlite to many times its original volume. Vermiculite is a form of mica with many ultra-thin layers between which are minute amounts of water, which expand or exfoliate the natural material to many times its original volume when subjected to high temperatures by the water turning into steam and forcing the layers of flakes apart. The density of a lightweight plaster is about one-third that of a comparable sanded plaster and it has a thermal value of about three times that of sanded plasters, resulting in a reduction of heat loss, less condensation and a reduction in the risk of pattern staining. They also have superior adhesion properties to all backgrounds including smooth dry clean concrete.

The choice of plaster mix, type and number of coats will depend upon the background or surface to which the plaster is to be applied. Roughness and suction properties are two of the major considerations. The roughness can affect the keying of the plaster to its background; special keyed bricks are available for this purpose (see Fig. 3.1.10), or alternatively the joints can be raked out to a depth of 15–20 mm as the wall is being built. *In-situ* concrete can be cast using sawn formwork giving a rough texture hence forming a suitable key. Generally all lightweight concrete blocks provide a suitable key for the direct application of plasters. Bonding agents in the form of resin emulsions are available for smooth surfaces – these must be applied in strict accordance with the manufacturer's instructions to achieve satisfactory results.

The suction properties of the background can affect the drying rate of the plaster by absorbing the moisture of the mix: too much suction can result in the plaster failing to set properly – thus losing its adhesion to its background; too little suction can give rise to drying shrinkage cracks due to the retention of excess water in the plaster.

Undercoat plasters are applied by means of a wooden float or rule worked between dots or runs of plaster to give a true and level surface. The runs or rules and dots are of the same mix as the backing coat and are positioned over the background at suitable intervals to an accurate level so that the main application of plaster can be worked around the guide points. The upper surface of the undercoat plaster should be scored or scratched to provide a suitable key for the finishing coat. The thin finishing coat of plaster is applied to a depth of approximately 3 mm and finished with a steel float to give a smooth surface. In hot and dry conditions care must be taken to ensure that water is not allowed to evaporate from the mix. Water is very important in the setting process and if applied plaster is allowed to dry too quickly, loss of strength, surface adhesion and surface irregularities will become apparent.

Most paints and wall coverings are compatible with plaster finishes, provided the plaster is thoroughly dried out prior to application. This normally takes several months; therefore, as an interim measure, permeable water-based emulsion paint could be used.

Guidance in application and standards of workmanship can be found in BS 5492: *Code of practice for internal plastering*, and BS 8000: Part 10: *Workmanship on building sites – Code of practice for plastering and rendering*.

DRY LINING TECHNIQUES

External walls or internal walls and partitions can be dry lined with a variety of materials which can be self finished, ready for direct decoration or have a surface suitable for a single final coat of board finish plaster. The main advantages of dry lining techniques are speed, reduction in the amount of water used in the construction of buildings thus reducing the drying out period and in some cases increased thermal insulation.

Suitable materials are hardboard, plywood, chipboard and plasterboard. Hardboard, plywood and chipboard are fixed to timber battens attached to the wall at centres to suit the spanning properties and module size of the board. Finishing can be a direct application of paint, varnish or wallpaper but masking the fixings and joints may present problems. As an alternative the joints can be made a feature of the design by the use of edge chamfers or by using moulded cover fillets.

Plasterboard consists of an aerated gypsum core encased in and bonded to specially prepared bonded paper liners. The grey-coloured liner is intended for a skim coat of plaster and the ivory-coloured liner is for direct decoration.

Gypsum plasterboards are manufactured in a variety of specifications to suit numerous applications. They are generally available in metric co-ordinated widths of 900 and 1,200 mm with co-ordinated lengths from 1,800 to 3,000 mm in thicknesses of 9.5 and 12.5 mm. Boards can be obtained with a tapered edge for a seamless joint or a square edge for cover fillet treatment. Some variants are listed in Table 7.8.2.

Table 7.8.2 Common types of plasterboard

Type	Thickness (mm)	Width (mm)	Application
Plank	19	600	Sound insulation
Moisture resistant	9.5 and 12.5	1,200	Bathroom/kitchen
Contoured	6	1,200	Curved surfaces
Vapour check	9.5 and 12.5	900 and 1,200	External wall/roof
Firecheck	12.5	900 and 1,200	Fire cladding
Lath	9.5 and 12.5	400 and 600	Ceiling lining
Thermal	12.5*	1,200	Thermal insulation
Glass reinforced	4–35	Various	Fire protection

* Overall thickness including laminates of expanded polystyrene, extruded polystyrene or phenolic foam, can be between 25 and 50 mm.

Figure 7.8.3   Plasterboard ceilings

Baseboard is produced with square edges as a suitable base for a single coat of board finish plaster. One standard width is manufactured with a thickness of 9.5 mm; lengths are 1,200, 1,219 and 1,372 mm. Before the plaster coat is applied the joints should be covered with a jute scrim 100 mm wide (see Fig. 7.8.3).

Plasterboard lath is a narrow plasterboard with rounded edges which removes the need for a jute scrim over the joints. The standard width is 406 mm with similar lengths to baseboard and thicknesses of 9.5 and 12.5 mm.

The fixing of dry linings is usually by nails to timber battens suitably spaced, plumbed and levelled to overcome any irregularities in the background. Fixing battens are placed vertically between the horizontal battens fixed at the junctions of the ceiling and floor with the wall. It is advisable to treat all fixing battens with an insecticide and a fungicide to lessen the risk of beetle infestation and fungal attack. The spacing of the battens will be governed by the spanning properties of the lining material; fixing battens at 450 mm centres are required for 9.5 mm thick plaster-boards and at 600 mm centres for 12.5 mm thick plasterboards, so placed that they coincide with the board joints.

12.5 mm plasterboard can also be fixed to brick or block masonry with dabs of plaster. Dabs of board finish plaster about 50–75 mm wide by 250 mm long are close spaced vertically at a horizontal spacing (max. 600 mm) to suit the board width. Intermediate dabs are similarly spaced with a continuous spread at ceiling and ground levels. The plasterboard is then placed in position, tapped and horizontally aligned with a spirit level until it is firmly in contact with the dabs. Double-headed (duplex) nails can be used to secure the boards temporarily while the dabs set, after which they are removed and joints are made good. It is recommended that tapered edge boards are used for this method of fixing.

GLAZED WALL TILES

Two processes are used in the manufacture of these hard glazed tiles. Firstly, the body or 'biscuit' tile is made from materials such as china clay, ball clay, flint and limestone which are mixed by careful processes into a fine powder before being heavily pressed into the required shape and size. The tile is then fired at a temperature of up to 1,150 °C. Secondly, glazing is applied in the form of a mixed liquid consisting of fine particles of glaze and water. The coated tiles are then fired for a second time at a temperature of approximately 1,050 °C when the glazing coating fuses to the surface of the tile. Gloss, matt and eggshell finishes are available together with a wide choice of colours, designs and patterns.

A range of fittings with rounded edges and mitres are produced to blend with the standard 150 × 150 mm square tiles, 5 or 6 mm thick; also a similar range of fittings are made for the 108 × 108 mm, 4 or 6 mm thick, square tiles. The appearance and easily cleaned surface of glazed tiles makes them suitable for the complete or partial tiling of bathrooms and the provision of splashbacks for sinks and basins.

Tiles are fixed with a suitable adhesive which can be of a thin bed of mastic adhesive or a bed of a cement-based adhesive; the former requires a flat surface such as a mortar screed. Ceramic or glazed tiles are considered to be practically inert, therefore careful selection of the right adhesive to suit the backing and final condition is essential. Manufacturer's instructions or the recommendations of BS 5385 should always be carefully followed.

Glazed tiles can be cut easily using the same method employed for glass. The tile is scored on its upper face with a glass cutter, this is followed by tapping the back of the tile behind the scored line over a rigid and sharp angle such as a flat straight edge.

CEILING FINISHES

Ceilings can be finished by any of the dry-lining techniques previously described for walls. The usual method is a plasterboard base with a skim coat of plaster. The plasterboards are secured to the underside of the floor or ceiling joists with galvanized plasterboard nails to reduce the risk of corrosion to the fixings. At the abutment of boards, a 3–5 mm gap is filled with jointing plaster and bridged with jute scrim or special jointing tape. The most vulnerable point in a ceiling to cracking is at the junction between the ceiling and wall; this junction should be strengthened with a jute scrim or jointing tape around the internal angle (see Fig. 7.8.3), or alternatively the junction can be masked with a decorative plasterboard or polystyrene cove moulding.

The cove moulding is made in a similar manner to plasterboard and is intended for direct decoration. Plasterboard cove moulding is jointed at internal and external angles with a mitred joint and with a butt joint in the running length. Any clean, dry and rigid background is suitable for the attachment of plasterboard cove which can be fixed by using a special water-mixed adhesive applied to the contact edges of the moulding which is pressed into position; any surplus adhesive should be removed from the edges before it sets. A typical plasterboard cove detail is shown in Fig. 7.8.3.

Many forms of ceiling tiles are available for application to a joisted ceiling or solid ceiling with a sheet or solid background. Fixing to joists should be by concealed or secret nailing through the tongued and grooved joint. If the background is solid such as a concrete slab then dabs of a recommended adhesive are used to secure the tiles. Materials available include expanded polystyrene, mineral fibre, fibreboard and glass fibre with a rigid vinyl face.

Other forms of finish which may be applied to ceilings are sprayed plasters which can be of a thick or thin coat variety. Spray plasters are usually of a proprietary mixture applied by spraying apparatus directly on to the soffit giving a coarse texture which can be trowelled smooth, patterned with combs or stippled with a stiff brush (artexing). Application can be with a brush if preferred. Various patterned ceiling papers are produced to give a textured finish. These papers are applied directly to the soffit or over a stout lining paper. Some ceiling papers are designed to be a self finish but others require one or more coats of emulsion paint.

INTERNAL FIXINGS AND SHELVES

INTERNAL FIXINGS

These consist of trims in the form of skirtings, dado rails, frieze or picture rails, architraves and cornices; whereas fittings would include such things as cupboards and shelves.

SKIRTINGS

A skirting is a horizontal member fixed around the skirt or base of a wall primarily to mask the junction between a wall finish and a floor. It can be an integral part of the floor finish such as quarry tile or PVC skirtings or it can be made from timber, metal or plastic. Timber is the most common material used and is fixed by nails direct to the background, but if it is a dense material that will not accept nails, plugs or special fixing bricks can be built into the wall. External angles in skirtings are formed by mitres but internal angles are usually scribed (see Fig. 7.9.1).

ARCHITRAVES

These are mouldings cut and fixed around door and window openings to mask the joint between the wall finish and the frame. Like skirtings the usual material is timber but metal and plastic mouldings are available. Architraves are fixed with nails to the frame or lining and to the wall in a similar manner to skirtings if the architrave section is large (see Fig. 7.9.1).

DADO RAILS

These are horizontal mouldings fixed in a position to prevent the walls from being damaged by the backs of chairs pushed against them. They are very seldom used today since modern chair design renders them unnecessary. If used, they are fixed by nails directly to the wall or to plugs inserted in the wall.

Position of picture rail if fitted

Wall finish

75 mm block partition wall

Mitred joint at head

Door lining

Planted door stop

70 × 20 chamfered and rounded architrave

Butt joint

95 × 20 bevel rounded skirting

Mitred joint

Moulding scribed to form mitre

Butt joint against vertical face of skirting

Scribed skirting forming internal angle joint

10

4

14

45

14

Typical picture rail moulding

Figure 7.9.1   Typical wood trim details

PICTURE RAILS

These are moulded rails fixed horizontally around the walls of a room from which pictures may be suspended and are usually positioned in line with the top edges of the door architrave. They can be of timber or metal and like the dado rail are very seldom used in modern domestic buildings. They would be fixed by nails in the same manner as dado rails and skirtings; a typical picture rail moulding is shown in Fig. 7.9.1.

CORNICES

Cornices are timber or plaster ornate mouldings used to mask the junction between the wall and ceiling. These are very seldom used today having been superseded by the cove mouldings.

CUPBOARD FITTINGS

These are usually supplied as a complete fitting and only require positioning on site; they can be free standing or plugged and screwed to the wall. Built-in cupboards can be formed by placing a cupboard front in the form of a frame across a recess and then hanging suitable doors to the frame. Another method of forming built-in cupboards is to use a recessed partition wall to serve as cupboard walls and room divider, and attach to this partition suitable cupboard fronts and doors.

SHELVES

Shelves can be part of a cupboard fitting or can be fixed to wall brackets or bearers which have been plugged and screwed to the wall. Timber is the usual material for shelving; this can be faced with a large variety of modern plastic finishes. Shelves are classed as solid or slat, the latter being made of 45×20 mm slats, spaced about 20 mm apart, and are used mainly in airing and drying cupboards; typical shelf details are shown in Fig. 7.9.2.

KITCHEN FITTINGS

Kitchen cupboards and drawer units are manufactured in preassembled or flat pack form. Flat pack is more convenient for storage and delivery, but requires time for site assembly before fitting. Materials vary, they may be selected timber or its less expensive derivatives and composites of plywood, chipboard or medium density fibreboard. Purpose-made plastic and metal brackets are used to assemble the components.

Disposition of fitments and general design layout is a specialist area of design for the application of standardized metric units shown in Fig. 7.9.3. Special components such as corner pieces and purpose-made base units, e.g. wine racks, are not included. These vary in style and measurements between manufacturers, but retain overall dimensions to co-ordinate with the standard range.

20 mm gap

45 × 20 slats fixed to bearers

45 × 20 bearer to each end fixed to wall

Solid shelf out of t & g boarding

45 × 20 bearer to each end fixed to wall

Solid chipboard shelf

45 × 20 back bearer

Rebated timber edging

45 × 20 bearer to each end fixed to wall

Solid blockboard shelf

Timber brackets fixed to wall

Typical pressed steel bracket

Figure 7.9.2 Shelves and supports

Figure 7.9.3 Standard kitchen units (dimensions in mm)

Heights of floor-supported units are nominal, with adjustable legs fitted as standard. Surfaces or worktops are secured to the base units with small angle brackets. The worktops are available in laminates of solid timber (usually beech), although the less expensive pattern and/or grain effect plastic-faced particle board is most often used. Thickness varies between 28 and 40 mm, with the option of a rounded edge or wood trimming.

IRONMONGERY

Ironmongery is a general term which is applied to builder's hardware and includes such items as nails, screws, bolts, hinges, locks, catches, window and door fittings.

NAILS

A nail is a fixing device which relies on the grip around its shank and the shear strength of its cross-section to fulfil its function. It is therefore important to select the right type and size of nail for any particular situation. Nails are specified by their type, length and diameter range given in millimetres. The diameter range is comparable to the old British Standard Wire Gauge. The complete range of nails is given in BS 1202. Steel is the main material used in the manufacture of nails; other metals used are copper and aluminium alloy.

Nails in common use are:

Cut clasp nails   These are made from black rolled steel and used for general carcassing work.

Cut floor brads   These are made from black rolled steel and used mainly for fixing floor boards, because their rectangular section reduces the tendency of thin boards splitting.

Round plain head   Also known as round wire nails, these are made in a wide variety of lengths. They are used mainly for general carpentry work but have a tendency to split thin members.

Oval brad head   These are made with an oval cross-section to lessen the risk of splitting the timber members. They are used for the same purpose as round wire nails but have the advantage of being able to be driven below the surface of the timber. A similar nail with a smaller head is also produced and is called an oval lost head nail.

Clout nails   Also called slate nails, these have a large flat head and are used for fixing tiles and slates. The felt nail is similar but has a larger head and is only available in lengths up to 38 mm.

Panel pins   These are fine wire nails with a small head which can be driven below the surface. They are used mainly for fixing plywood and hardboard.

Plasterboard nails   The holding power of these nails, with their countersunk head and jagged shank, makes them suitable for fixing ceiling and similar boards.

WOOD SCREWS

A wood screw is a fixing device used mainly in joinery and relies upon its length and thread for holding power and resistance to direct extraction. For a screw to function properly it must be inserted by rotation and not be driven in with a hammer. It is usually necessary to drill pilot holes for the shank and/or core of the screw.

Wood screws are manufactured from cold drawn wire, steel, brass, stainless steel, aluminium alloy, silica bronze and nickel–copper alloy. In addition to the many different materials a wide range of painted and plated finishes are available such as an enamelled finish known as black japanned. Plated screws are used mainly to match the fitting which they are being used to fix and include such finishes as galvanized steel, copper plated, nickel plated and bronze metal antique.

Screws are specified by their material, type, length and gauge. The screw gauge is the diameter of the shank and is designated by a number; but, unlike the gauge used for nails, the larger the screw gauge number the greater the diameter of the shank. Various head designs are available for all types of wood screws, each having a specific purpose:

Countersunk head   This is the most common form of screw head and is used for a flush finish, the receiving component being countersunk to receive the screw head.

Raised head   This is used mainly with good quality fixtures, the rounded portion of the head with its mill slot remains above the surface of the fixture ensuring that the driving tool does not come into contact with the surface causing damage to the finish.

Round head   The head, being fully exposed, makes these screws suitable for fittings of material which is too thin to be countersunk.

Recessed head   Screws with a countersunk head and a recessed cruciform slot giving a positive drive with a specially shaped screwdriver.

Coach screws   These are made of mild steel with a square head for driving in with the aid of a spanner and used mainly for heavy carpentry work.

CAVITY FIXINGS

Various fixing devices are available for fixing components to thin materials of low structural strength such as plasterboard and hardboard. Cavity fixings are designed to spread the load over a wide area of the board, typical examples are:

Steel spring toggles   Spring-actuated wings open out when the toggle fixing has been inserted through a hole in the board and spread out on the reverse side of the board. Spring toggles are specially suited to suspend fixtures from a ceiling.

Steel gravity toggles   When inserted horizontally into a hole in the board the long end of the toggle drops and is pulled against the reverse side of the board when the screw is tightened.

Rubber cavity fixings   A rubber or neoprene sleeve, in which a nut is embedded, is inserted horizontally through a hole in the board, the tightening of the screw causing the sleeve to compress and grip the reverse side of the board. This fixing device forms an airtight, waterproof and vibration resistant fixing.

Typical examples of nails, screws and cavity fixings are shown in Fig. 7.10.1.

HINGES

Hinges are devices used to attach doors, windows and gates to a frame, lining or post so that they are able to pivot about one edge. It is of the utmost importance to specify and use the correct number and type of hinge in any particular situation to ensure correct operation of the door, window or gate. Hinges are classified by their function, length of flap, material used and sometimes by the method of manufacture. Materials used for hinges are steel, brass, cast iron, aluminium and nylon with metal pins. Typical examples of hinges in common use are:

Steel butt hinge   These are the most common type in general use and are made from steel strip which is cut to form the flaps and is pressed around a steel pin.

Steel double flap butt hinge   Similar to the steel butt hinge but made from two steel strips to give extra strength.

Rising butt hinge   This is used to make the door level rise as it is opened to clear carpets and similar floor coverings. The door will also act as a gravity self closing door when fitted with these butts which are sometimes called skew butt hinges.

Parliament hinge   This is a form of butt hinge with a projecting knuckle and pin enabling the door to swing through 180° to clear architraves and narrow reveals.

Tee hinge   Sometimes called a cross garnet, these hinges are used mainly for hanging matchboarded doors where the weight is distributed over a large area.

Band and hook   A stronger type of tee hinge made from wrought steel and used for heavy doors and gates. A similar hinge is produced with a pin which projects from the top and bottom of the band and is secured with two retaining cups screwed to the post; these are called reversible hinges.

Typical examples of common hinges are shown in Fig. 7.10.2.

Figure 7.10.1   Nails, screws and cavity fixings

Figure 7.10.2 Typical hinges

LOCKS AND LATCHES

Any device used to keep a door in the closed position can be classed as a lock or latch. A lock is activated by means of a key whereas a latch is operated by a lever or bar. Latches used on lightweight cupboard doors are usually referred to as catches. Locks can be obtained with a latch bolt so that the door can be kept in the closed position without using a key; these are known as deadlocks.

Locks and latches are either fixed to the face of the door with a staple or keep fixed to the frame when they are termed rim locks or latches. If they are fixed within the body of the door they are called mortice locks or latches. When this form of lock or latch is used the bolts are retained in mortices cut behind the striking plate fixed to the frame (see Fig. 7.10.3).

Cylindrical night latches are fitted to the stile of a door and a connecting bar withdraws the latch when the key is turned. Most night latches have an internal device to stop the bolt being activated from outside by the use of a key.

Door handles, cover plates and axles used to complete lock or latch fittings are collectively called door furniture and are supplied in a wide range of patterns and materials.

DOOR BOLTS

Door bolts are security devices fixed to the inside faces of doors and consist of a slide or bolt operated by hand to locate in a keep fixed to the frame. Two general patterns are produced: the tower bolt, which is the cheapest form; and the stronger, but dearer, barrel bolt. The bolt of a tower bolt is retained with staples or straps along its length, whereas in a barrel bolt it is completely enclosed along its length (see Fig. 7.10.3).

CASEMENT WINDOW FURNITURE

Two fittings are required for opening sashes, the fastener, which is the security device, and the stay, which holds the sash in the opened position. Fasteners operate by the blade being secured in a mortice cut into the frame or by the blade locating over a projecting wedge or pin fixed to the frame (see Fig. 7.10.3). Casement stays can be obtained to hold the sash open in a number of set positions by using a pin fixed to the frame and having a series of locating holes in the stay or they can be fully adjustable by the stay sliding through a screw down stop fixed to the frame (see Fig. 7.10.3).

LETTER PLATES

These are the hinged covers attached to the outside of a door, which cover the opening made to enable letters to be delivered through the door. The minimum

Staple fixed to frame

Latch bolt

Case

Lock bolt

Rim lock

Striking plate fixed to frame

Latch bolt

Case

Lock bolt — Forend

Mortice lock

Backplate

Lever handle

Axle

Wedge plate fixed to frame

Backplate

Lever mortice furniture

Fastener fixed to sash

Casement fastener

Stay fixed to sash — Pin fixed to frame

Casement stay

Staple fixed to frame

Straps

Bolt

Plate fixed to door

Tower bolt

Staple fixed to frame

Barrel

Bolt

Plate fixed to door

Barrel bolt

Figure 7.10.3   Door and window ironmongery

opening size for letter plates recommended by the postal services is 200×45 mm; the bottom of the opening should be sited not lower than 750 mm from the bottom edge of the door and not higher than 1,450 mm to the upper edge of the opening. A wide range of designs are available in steel, aluminium alloy and plastic, some of which have a postal knocker incorporated in the face design (see BS 2911).

BS 4951 and BS 5872 cover builder's hardware for housing and give recommendations for materials, finishes and dimensions to a wide range of ironmongery items not covered by a specific standard. These include such fittings as finger plates, cabin hooks, gate latches, cupboard catches and drawer pulls.

PAINTING AND DECORATING

The application of coats of paint to the elements, components, trims and fittings of a building has two functions. The paint will impart colour and, at the same time, provide a protective coating which will increase the durability of the member to which it has been applied. The covering of wall and ceiling surfaces with paper or fabric is basically to provide colour, contrast and atmosphere. To achieve a good durable finish with any form of decoration the preparation of the surface and the correct application of the paint or paper is of the utmost importance.

PAINT

Paint is a mixture of a liquid or medium and a colouring or pigment. Mediums used in paint manufacture range from thin liquids to stiff jellies and can be composed of linseed oil, drying oils, synthetic resins and water. The various combinations of these materials form the type of class of paint. The medium's function is to provide the means of spreading and binding the pigment over the surface to be painted. The pigment provides the body, colour and durability of the paint. White lead is a pigment which gives good durability and moisture resistance but it is poisonous, therefore its use is confined mainly to priming and undercoating paints in situations where it is not exposed to human contact. Paints containing a lead pigment are required by law to state this fact on the can. The general pigment used for finishing paint is titanium dioxide which gives good obliteration of the undercoating but is not poisonous.

OIL-BASED PAINTS

Priming paints   These are first coat paints used to seal the surface, protect the surface against damp air, act as a barrier to prevent any chemical action between the surface and the finishing coats and give a smooth surface for the subsequent coats. Priming paints are produced for application to wood, metal and plastered surfaces.

356

Undercoating paints   These are used to build up the protective coating and to provide the correct surface for the finishing coat(s). Undercoat paints contain a greater percentage of pigment than finishing paints and as a result have a matt or flat finish. To obtain a good finishing colour it is essential to use an undercoat of the type and colour recommended by the manufacturer.

Finishing paints   A wide range of colours and finishes including matt, semi-matt, eggshell, satin, gloss and enamel are available. These paints usually contain a synthetic resin which enables them to be easily applied, quick drying and have good adhesive properties. Gloss paints have less pigment than the matt finishes and consequently less obliterating power.

POLYURETHANE PAINTS

These are quick-drying paints based on polyurethane resins giving a hard heat-resisting surface. They can be used on timber surfaces as a primer and undercoat but metal surfaces will require a base coat of metal primer; the matt finish with its higher pigment content is best for this 'one paint for all coats' treatment. Other finishes available are gloss and eggshell.

WATER-BASED PAINTS

Most of the water-based paints in general use come under a general classification of emulsion paints: they are quick drying and can be obtained in matt, eggshell, semi-gloss and gloss finishes. The water medium has additives such as polyvinyl acetate and alkyd resin to produce the various finishes. Except for application to iron work, which must be primed with a metal primer, emulsion paints can be used for priming, undercoating and as a finishing application. Their general use is for large flat areas such as ceilings and walls.

VARNISHES AND STAINS

Varnishes form a clear, glossy or matt, tough film over a surface and are a solution of resin and oil, their application being similar to oil based paints. The type of resin used, together with the correct ratio of oil content, forms the various durabilities and finishes available.

Stains can be used to colour or tone the surface of timber before applying a clear coat of varnish; they are basically a dye in a spirit and are therefore quick drying.

PAINT SUPPLY

Paints are supplied in metal and plastic containers of 5 litres, 2.5 litres, 1 litre, 500 millilitres and 250 millilitres capacity. They are usually in one of the

100 colours recommended for building purposes in BS 4800. BS 381C is also a useful reference, giving 91 standard colours for special purposes such as determining colours used by HM services. The colour range in BS 381C can be used for purposes other than paint identification.

The British Standard colour range is derived from the Munsell scale of colours and includes black and white. The Munsell system is used internationally to provide an accurate definition of colours. It is based on hue (base colour), value (light or darkness) and chroma (intensity compared to greyness).

KNOTTING

Knots or streaks in softwood timber may exude resins which may soften and discolour paint finishes; generally an effective barrier is provided by two coats of knotting applied before the priming coats of paint. Knotting is a uniform dispersion of shellac in a natural or synthetic resin in a suitable solvent. It should be noted that the higher the grade of timber specified, with its lower proportion of knots, the lower will be the risk of resin disfigurement of paint finishes.

WALLPAPERS

Most wallpapers consist of a base paper of 80% mechanical wood pulp and 20% unbleached sulphite pulp printed on one side with a colour and/or design. Papers with vinyl plastic coatings making them washable are also available, together with a wide range of fabrics suitably coated with vinyl to give various textures and finishes for the decoration of walls and ceilings.

The preparation of the surface to receive wallpaper is very important; ideally a smooth clean and consistent surface is required. Walls which are poor can be lined with a light-weight lining paper for traditional wallpaper and a heavier esparto lining paper for the heavier classes of wallpaper. Lining papers should be applied in the opposite direction to the direction of the final paper covering.

Wallpapers are attached to the surface by means of an adhesive such as starch/flour paste and cellulose pastes which are mixed with water to the manufacturer's recommendations. Heavy washable papers and woven fabrics should always be hung with an adhesive recommended by the paper manufacturer. New plaster and lined surfaces should be treated to provide the necessary 'slip' required for successful paper hanging. A glue size is usually suitable for a starch/flour paste, whereas a diluted cellulose paste should always be used when a cellulose paste is used as the adhesive.

Standard wallpapers are supplied in rolls with a printed width of 530 mm and a length of 10 m giving an approximate coverage of 4.5 m^2 – the actual coverage will be governed by the 'drop' of the pattern which may be as much as 2 m. Decorative borders and motifs are also available for use in conjunction with standard wall-papers.

TYPES OF WALLPAPER

Underlayers

Lining paper   This is used to provide uniform suction on uneven surfaces and to reinforce cracked plaster before applying decorative wallpaper finishes. The quality varies from common pulp for normal use to a stronger brown paper for badly cracked surfaces. A pitch impregnated lining paper is also available for damp walls, but it is not intended as a substitute for remedial treatment to a source of dampness. Cotton-backed paper is produced for use on boarded backgrounds that are likely to move. Lining papers are manufactured 11 m long by 560 or 760 mm in width.

Finishes

Flocks   These are manufactured with a raised pile pattern produced by blowing coloured wool, cotton or nylon fibres onto an adhesive printed pattern.

Metallic finishes   Paper coated with metal powders or a metal foil sheet material.

Moires   A watered silk effect finish.

Ingrained   An effect produced by incorporating sawdust or other fibres in the paper surface (wood-chip).

Textured effect   An imitation texture achieved by embossing the surface.

Washable papers   These include varnish and plastic-coated paper, in addition to expanded PVC foam and textured PVC sheet.

Relief papers   These are in two categories – high relief (anaglypta) and low relief (lincrusta). Anaglypta is a quality wallpaper moulded while wet to provide embossed patterns which give the paper strength and resilience. The depressions on the back of the paper can be filled with a paste/sawdust mix to resist impact damage. Lincrusta is a paper composed of whiting, wood flour, lithopone, wax, resin and linseed oil which is superimposed hot on kraft paper. The result is a relatively heavy paper requiring a dense paste for adhesion to the wall.

PART 8

INSULATION

SOUND INSULATION

THE DEFINITION OF SOUND

Anything that can be heard is a sound, whether it is made by conversation, machinery, or walking on a hard surface. All sounds are produced by a vibrating object which moves rapidly to and fro causing movement of the tiny particles of air surrounding the vibrating source. The displaced air particles collide with adjacent particles setting them in motion and in unison with the vibrating object. Air particles move only to and fro but the sound wave produced travels through the air until at some distance from the source the movement of the particles is so slight that the sound produced is inaudible.

For a sound to be transmitted over a distance a substance, called the *sound medium*, is required. It can be shown by experiments that sound cannot travel through a vacuum but it can be transmitted through solids and liquids.

Sounds can differ in two important ways, by loudness and by pitch. The loudness of a sound depends on the distance through which the vibrating object moves to and fro as it vibrates; the greater the movement the louder the sound. The loudness with which a sound is heard depends upon how far away from the source the receiver or ear is. The unit of subjective loudness is a *phon* while the objective unit is called a *decibel*. Although the loudness of a sound will vary with the frequency of the note for practical building purposes, the phon and the decibel are considered to be equal over the range of 0 phons, the threshold of hearing, to 130 phons, the threshold of painful hearing.

The pitch of a sound depends on the rate at which the vibrating object oscillates. The number of vibrations in a second is called the *frequency* and the higher the frequency the higher the pitch. The lowest pitched note that the human ear can hear has a frequency of approximately 16 hertz whereas the highest pitched note which can be heard by the human ear has a frequency of approximately 20,000 hertz or cycles per second.

When a sound is produced within a building three reactions can occur:

1. The pressure or sound waves can come into contact with the walls, floor and ceiling and be reflected back into the building.

2. Some of the sound can be absorbed by these surfaces and/or the furnishes. It must be noted that sound absorption normally only benefits the occupants of the room in which the sound is generated since its function is to reduce the amount of reflected sound.
3. The sound waves upon reaching the walls, floor and ceiling can set these members vibrating in unison and thus transmit the sound to adjacent rooms.

It must also be noted that sounds can enter a building from external sources such as traffic and low-flying aircraft (see Fig. 8.1.1).

Sounds may be defined as either *impact* sounds, caused by direct contact with the structure such as footsteps and hammering on walls which will set that part of the structure vibrating, or they can be termed *airborne* sounds, such as the conversation or radio which sets the structure vibrating only when the sound waves emitted from the source reach the structural enclosure.

A noise can be defined as any undesired sound and may have any one of the following effects:

■ Annoyance.
■ Disturbance of sleep.
■ Interfere with the ability to hold a normal conversation.
■ Damage to hearing.

It is difficult to measure annoyance since it is a subjective attitude and will depend upon the mental and physical well being of the listener, together with the experience of being subjected to such types of noise. Damage to hearing can be caused by a sudden noise such as a loud explosion or by gradual damage resulting from continual noise over a period of years.

The solution to noise or sound problems can only therefore be reasonable to cater for the average person and conditions. The approach to solving a noise problem can be three-fold:

1. Reduce the noise emitted at the source by such devices as attenuators and mounting machinery on resilient pads.
2. Provide a reasonable degree of sound insulation to reduce the amount of sound transmitted.
3. Isolate the source and the receiver.

SOUND INSULATION

The most effective barrier to the passage of sound is a material of high mass. With modern materials and methods this form of construction is both impracticable and uneconomic. Unfortunately modern living with its methods of transportation and entertainment generates a considerable volume of noise and therefore some degree of sound insulation in most buildings is not only desirable but mandatory under building law.

Figure 8.1.1   Sources of noise within and around buildings

BUILDING REGULATIONS

Parts E1 to E3 of the Building Regulations considers sound insulation to all types of dwelling. This includes hostels, hotels, student accommodation, nurses' homes and homes for the elderly, but excludes hospitals and prisons. Part E4 is specifically to control acoustic conditions in new schools, i.e. between rooms, from external noise and to create appropriate sound reverberation characteristics for clarity of speech.

The Approved Documents supporting these regulations are as follows:

- E1 Protection against sound from adjoining dwellings or buildings etc.
- E2 Protection against sound within a dwelling etc.
- E3 Reverberation in the common internal parts of buildings containing dwellings etc.
- E4 Acoustic conditions in schools.

PROTECTION AGAINST SOUND FROM ADJOINING DWELLINGS OR BUILDINGS – (FIGS. 8.1.2 AND 8.1.3)

This will be satisfied with separating elements of structure (walls, floors and stairs between dwellings and residential buildings) having the following minimum standards of sound insulation:

- Dwellings purpose built –
  Airborne sound 45 dB, impact sound (floors/stairs) 62 dB.
- Dwellings created by change of use –
  Airborne sound 43 dB, impact sound (floors/stairs) 64 dB.
- Residential purpose built –
  Airborne sound (walls) 43 dB, impact sound (walls) n/a.
  Airborne sound (floors/stairs) 45 dB, impact sound (floors/stairs) 62 dB.
- Residential created by change of use –
  Airborne sound (walls) 43 dB, impact sound (walls) n/a.
  Airborne sound (floors/stairs) 43 dB, impact sound (floors/stairs) 64 dB.

PROTECTION AGAINST SOUND WITHIN A DWELLING

This applies specifically to walls (it should also apply to floors) between a WC compartment and habitable rooms; walls between bedrooms, bedrooms and other rooms and floors between bedrooms and other bedrooms and other rooms. Performance is satisfied if the minimum airborne sound insulation between these walls, floors and stairs is 40 dB. This is verified by types of construction assembled, tested and measured in laboratory controlled conditions. Examples include:

Walls –
- Timber (75 mm min.) or metal stud (45 mm min.) framing with two layers of plasterboard (min. mass 9 kg/m^2 per sheet) each side.
- Concrete blockwork (min. mass 120 kg/m^2) with plaster or plasterboard finish both sides.

Floors –

■ Timber joists with wood board (min. mass 15 kg/m²) upper surface, plasterboard (min. mass 9 kg/m²) ceiling and 100 mm mineral wool (min. density 10 kg/m³) in void.

■ 40 mm min. screed over precast concrete beams and infill blocks (min. mass exc. screed 220 kg/m²) with timber battens and plasterboard (min. mass 9 kg/m²) soffit.

Figure 8.1.2 Separating walls

Figure 8.1.3   Separating floors

PROTECTION AGAINST NOISE FROM EXTERNAL SOURCES

This requirement is generally satisfied where external elements are constructed in accordance with structural and thermal insulation requirements (Building Regulations A and L respectively). For example, brick and block cavity masonry or brick clad timber framed walls, tiled or slated roof containing insulation and 12.5 mm plasterboard ceiling, double glazed windows with trickle vents (see additional section headed EXTERNAL NOISE on this page).

REVERBERATION IN THE COMMON INTERNAL PARTS OF BUILDINGS CONTAINING DWELLINGS

This is satisfied by applying absorptive materials, e.g. carpet, as a surface finish. Different applications, i.e. entrance halls, corridors, etc. will have suitable coefficients of sound absorption depending on the building purpose. These are listed in the Approved Document to the Building Regulations and correspond to a range of octave frequencies. Worked examples are also shown.

ACOUSTIC CONDITIONS IN SCHOOLS

Standards of construction should be adequate to satisfy guidelines for sound insulation, reverberation time and internal noise as determined in the publications:

Building Bulletin 87 – Guidelines for Environmental Design in Schools, and Building Bulletin 93 – The Acoustic Design of Schools.

Both are produced by the DfEE and published by the Stationery Office.

[*Note*: Where the local building control authority or other control body are required to determine conformity to standards, random sample testing of some houses on an estate development is deemed adequate. Site equipment is a simple hand held meter which can be used to indicate the sound level from an established noise source on one side of a separating element and that occurring on the other. Variations in air pressure are converted to a metered scale in volts which are depicted in units of decibels as the *root mean square (RMS)* of the sound. The *RMS* of sound is a type of uniform average.]

EXTERNAL NOISE

The main barrier to external noise is provided by the shell or envelope of the building, the three main factors being:

1. The mass of the enclosing structure.
2. The continuity of the structure.
3. Isolation by double leaf construction when using lightweight materials.

Generally the main problem for the insulation against external noise is the windows, particularly if these can be opened for ventilation purposes. Windows cannot provide the dual function of insulation against noise and ventilation, since the admission of air will also admit noise. Any type of window when opened will give a sound reduction of about 10 decibels as opposed to the 45–50 decibel reduction of the traditional cavity wall. A closed window containing single glazing will give a reduction of about 20 decibels, or approximately half that of the surrounding wall. It is obvious that the window to wall ratio will affect the overall sound reduction of the enclosing structure.

Double glazing can greatly improve the sound insulation properties of windows provided the following points are observed:

■ Sound insulation increases with the distance between the glazed units; for a reduction of 40 decibels the airspace should be 150–200 mm wide.
■ If the double windows are capable of being opened they should be weather-stripped.
■ Sound insulation increases with glass thickness particularly if the windows are fixed; this may mean the use of special ventilators having specific performances for ventilation and acoustics.

Double glazing, designed specifically to improve the thermal properties of a window, has no real value for sound insulation except by increasing the mass of resistance.

Roofs of traditional construction and of reinforced concrete generally give an acceptable level of sound insulation, but the inclusion of rooflights can affect the resistance provided by the general roof structure. Lightweight roofing such as profiled steel or fibre cement will provide only a 15–20 decibel reduction, but is generally acceptable on industrial buildings where noise is generated internally by the manufacturing process. Furthermore, the mandatory inclusion of thermal insulation in the roof slope will benefit sound reduction by its inherent absorbing properties. The inclusion of rooflights in this type of roof generally has no adverse effects in an uninsulated roof, since the sound insulation values of the rooflight materials are similar to those of the coverings. However, this is not so if the main elements of the roof are insulated.

Modern buildings can be designed to give reasonable sound insulation and consequent comfort to the occupiers, but improvements and upgrading to existing properties can present practical problems. In addition to the Building Regulations, Approved Documents E1 to E5, the following provide a useful source of information on reducing noise in existing buildings:

■ BS 8233: *Code of practice for sound insulation and noise reduction for buildings.*
■ BRE Report 238: *Sound control for homes.*
■ BRE Digest 293: *Improving the sound insulation of separating walls and floors* (see also Digests 333 and 334).
■ BRE Information Paper 6/88: *Methods for improving the sound insulation between converted flats.*

THERMAL INSULATION

Thermal insulation may be defined as a barrier to the natural flow of heat from an area of high temperature to an area of low temperature. In buildings this flow is generally from the interior to the exterior. Heat is a form of energy consisting of the ceaseless movement of tiny particles of matter called molecules; if these particles are moving fast they collide frequently with one another and the substance becomes hot. Temperature is the measure of hotness and should not be confused with heat.

The transfer of heat can occur in three ways:

- Conduction – vibrating molecules come into contact with adjoining molecules and set them vibrating faster and hence they become hotter; this process is carried on throughout the substance without appreciable displacement of the particles.
- Convection – transmission of heat within a gas or fluid caused by the movement of particles which become less dense when heated and rise thus setting up a current or circulation.
- Radiation – heat is considered to be transmitted by radiation when it passes from one point to another without raising the temperature of the medium through which it travels.

In a building all three methods of heat transfer can take place since the heat will be conducted through the fabric of the building and dissipated on the external surface by convection and/or radiation.

The traditional thick and solid building materials used in the past had a natural resistance to the passage of heat in large quantities, whereas the lighter and thinner materials used today generally have a low resistance to the transfer of heat. Therefore to maintain a comfortable and healthy internal temperature the external fabric of a building must be constructed of a combination of materials which will provide an adequate barrier to the transfer of heat.

Thermal insulation of buildings will give the following advantages:

- Reduction in the rate of heat loss.
- Lower capital costs for heating equipment.
- Lower fuel costs.
- Reduction in the risk of pattern staining.

- Reduction of condensation and draughts thus improving the comfort of the occupants.
- Less maintenance and replacement costs of heating equipment.
- Conservation of diminishing fossil fuel reserves.
- Reduction of atmospheric pollution, greenhouse effect and ozone depletion.

BUILDING REGULATIONS

Building Regulation L1 states that reasonable provision shall be made for the conservation of fuel and power in dwellings. The requirements of this regulation can be satisfied by limiting the areas of windows, doors and rooflights and by not exceeding the maximum U values for elements which are given in Approved Document L1. In addition to giving the maximum thermal transmittance coefficient or U value for various situations, the document also provides alternative formula methods for satisfying the Regulations. U values are expressed in $W/m^2 K$, which is the rate of heat transfer in watts (joules/sec.) through $1 m^2$ of the structure for one unit of temperature difference between the air on the two sides of the structure.

To calculate a U value the complete constructional detail must be known, together with the following thermal properties of the materials and voids involved:

Thermal conductivity   This is represented by the symbol λ (lambda). It is the measure of a material's ability to transmit heat and is expressed as the energy flow in watts per square metre of surface area for a temperature gradient of one Kelvin per metre thickness, i.e. $W/m K$.

Thermal resistance   This is symbolized by the letter R, as representative of a material's thermal resistance achieved by dividing its thickness in metres by its thermal conductivity, i.e. $R = m/\lambda$ or $m^2 K/W$.

Surface or standard resistances   These are values for surface and air space (cavity) resistances. They vary with direction of energy flow, building elevation, surface emissivity and degree of exposure. They are expressed in the same units as thermal resistance. In calculations they are represented by:

- R_{si} = surface resistance inside (typical value, $0.12 m^2 K/W$)
- R_{so} = surface resistance outside (typical value, $0.06 m^2 K/W$)
- R_a = surface resistance of the air space (typical value, $0.18 m^2 K/W$)

[*Note*: Values for λ and R can be obtained by reference to tables in Approved Document L to the Building Regulations, tables published by the Chartered Institution of Building Services Engineers or from listings in product manufacturer's catalogues.]

The method for calculating the U value of any combination of materials will depend on the construction. For example, a traditional rendered and plastered one

Figure 8.2.1 Calculation of U value through an external insulated cavity wall

brick wall is consistent, therefore the U value is expressed as the reciprocal of the summation of resistances, i.e. $U = 1 \div \Sigma R$. ΣR comprises internal surface resistance (R_{si}), thermal resistance of materials (R_1 = plaster, R_2 = brickwork [inc. mortar] and R_3 = render) and external surface resistance (R_{so}). If an element of structure is inconsistent, e.g. lightweight concrete blockwork with dense mortar joints, or timber framing with insulation infilling, cold bridging will occur through the denser parts. The Building Regulations require this to be incorporated into the calculation using the Combined Method from BS EN ISO 6946: *Building components and building elements – Thermal resistance and thermal transmittance – Calculation method.*

Cavity wall U value calculation (ref. Fig. 8.2.1)

■ Butterfly pattern twisted wire wall ties can be used where the gap between masonry leaves does not exceed 75 mm. With these ties no allowance for thermal conductivity or bridging is considered necessary. In Figure 8.2.1 where the gap exceeds 75 mm, vertical twist pattern ties are generally specified and an allowance of 0.02 W/m² K should be added to the U value calculation.
■ As the thermal resistance of brick and mortar is similar, i.e. < 0.1 m² K/W, the outer leaf is considered consistent and unbridged.
■ Fractional area of bridged layer:

Standard block nominal area = 450×225 = 101,250 mm²
Standard block format area = 440×215 = 94,600 mm²
Difference or mortar area =   =   6,650 mm²

Block format = 94,600 ÷ 101,250 = 0.934 or 93.4%
Mortar area = 6,650 ÷ 101,250 = 0.066 or 6.6%

■ The example shown has only one bridged layer, the mortar joints in the inner leaf of brickwork.

Material	*Thermal resistance (R)*
Internal surface	0.120
Plaster	0.081
Blockwork – mortar (6.6%)	0.188
Blockwork – blocks (93.4%)	1.364
Insulation	1.818
Cavity	0.180
Brickwork (inc. mortar)	0.121
External surface	0.060
	3.932 m² K/W

Upper resistance limit:

Heat energy loss has two paths in the wall, through the blocks and through the mortar.

1. Resistance through blocks, $3.932 - 0.188 = 3.744$ m² K/W [R_1] Fractional area, 93.4% or 0.934 [F_1]
2. Resistance through mortar, $3.932 - 1.364 = 2.568$ m² K/W [R_2] Fractional area, 6.6% or 0.066 [F_2]

$$\text{Upper limit of resistance} = \frac{1}{\dfrac{F_1}{R_1} + \dfrac{F_2}{R_2}} = \frac{1}{\dfrac{0.934}{3.744} + \dfrac{0.066}{2.568}}$$

$$= 3.634 \text{ m}^2 \text{ K/W}$$

Lower limit of resistance is a summation of resistance of all parts of the wall:

Internal surface	= 0.120
Plaster	= 0.081
Blockwork $1 \div [(0.934 \div 1.364) + (0.066 \div 0.188)]$	= 0.965
Insulation	= 1.818
Cavity	= 0.180
Brickwork	= 0.121
External surface	= 0.060
	3.345 m² K/W

The total resistance of the wall is the average of upper and lower resistances:

$(3.634 + 3.345) \div 2 = 3.489$ m² K/W

U value $= 1 \div R = 1 \div 3.489 = 0.286$ W/m² K

To this value add, 0.020 (wall tie allowance) $= 0.306$ W/m² K

Further worked examples for various elements of structure, construction method and materials are shown in Approved Document L to the Building Regulations. The thermal insulation of roofs can be effected between and/or over the rafters beneath the covering or at ceiling level. Generally rafter level insulation will use

$U = 0.16$

$U = 0.20$

Windows, doors and rooflights, max. 25% of the total floor area

Pitched roof

Flat roof
$U = 0.25$

Exposed walls
$U = 0.35$

Exposed floors
$U = 0.25$

$U = 0.25$

$U = 0.35$

Ground floor
$U = 0.25$

Garage (unheated)

Area weighted average* of wood or PVC framed windows, doors and rooflights, $U = 2.00$. For metal frames, $U = 2.20$.
(*Depends on glass and door component individual U values)

SEDBUK minimum values: Mains gas – 78%
LPG   – 80%
Oil   – 85%

Maximum permissible elemental U values for dwellings

Top and sides of cold water storage cistern to be insulated

Insulation – for thickness required see Appendix A in Approved Document L

Baffle to provide continuous air gap over insulation

Low-density block cavity closure to avoid cold bridge (see also Fig. 6.2.7)

Rising warm air from below prevents freezing

Ventilation holes or slots equivalent to continuous 10 mm gap fitted with insect mesh

Figure 8.2.2 Building Regulation thermal insulation requirements more materials but can be applied as a combined roofing felt and insulation to economise in application costs. The roof void will be warm and on sheltered sites it should not be necessary to protect the water cistern and pipework against frost. Applying the insulation at ceiling level is simpler, but as the roof void is unheated the plumbing housed within this space will require insulation from frost damage (see Fig. 8.2.2). Cold roofs will need ventilation to prevent condensation in accordance with Building Regulation F2 and BS 5250: *Code of practice for control of condensation in buildings.*

ENERGY RATING OF DWELLINGS

The Government has established a Standard Assessment Procedure (SAP) for determining an energy rating for all new dwellings, including those created from a change of use of an existing structure, e.g. house or other conversions to flats. The SAP incorporates a number of considerations, including incidental solar gains, ventilation characteristics, type of water heater, energy control system/accessories, fuel prices and type of thermal insulation applied to a building.

The energy efficiency of a dwelling can be rated anywhere on a scale from 1 to 120, with figures in excess of 80 expected of new dwellings. Its precise evaluation is calculated by extracting data from tables and compiling worksheets. See Building Regulations, Approved Document L, Appendices G and M, and the "Government's Standard Assessment Procedure for Energy Rating of Dwellings – SAP 2001". There is no obligation to achieve a particular rating, but its value is to provide propective purchasers and tenants with a facility to compare energy use in dwellings and a means to estimate annual fuel costs.

Further compliance with the Building Regulations is by satisfying one of the following provisions for limiting heat losses through the building fabric:

- Elemental method
- Target *U* value method
- Carbon index method

Elemental method   This is satisfied if the thermal performance of the elements of construction conform to the criteria given in Fig. 8.2.2. Some possible applications include:

- Double glazing to all windows and glazed doors.
- External masonry wall of cavity construction, full or part filled with insulation as required.
- Timber framed external wall with brick outer leaf, 10 mm plywood cladding to frame, 140 mm mineral wool insulation between frame and a 25 mm plasterboard internal lining.
- Ground floor fully insulated, typically with 50 mm high compressive strength mineral wool or EPS board below reinforced screed or concrete.
- Roof insulation between and over ceiling joists with at least 200 mm mineral wool.

For detailed examples see Appendix A in Approved Document L to the Building Regulations and insulation product manufacturer's catalogues, Note: SEDBUK (ref. Fig. 8.2.2) is the acronym for Seasonal Efficiency of a Domestic Boiler in the United Kingdom. Every boiler is listed on a database developed by the Government's Energy Efficiency Best Practice programme, which provides a fair comparison of different manufacturers' models. Information can be found on the website, www.boilers.org.uk.

Target U value The elemental method can be restrictive of design potential, whereas the target U value method provides a more flexible assessment. It makes use of incidentals such as variable glazed areas to take advantage of solar gains, sophisticated energy management systems to control the internal environment and associated high efficiency water heaters, e.g. condensing boilers. Calculations are necessary to justify conformity and they must show that the target U value exceeds the average U value.

Formula:

$$U = 0.35 - [0.19(A_R/A_T) - 0.10(A_{GF}/A_T) + 0.413(A_F/A_T)]$$

where: A_R = Area of exposed roof
A_T = Total area of exposed elements
A_{GF} = Area of ground floor
A_F = Total area of all floors (all storeys)

See Building Regulations, Approved Document L, Appendix F for application and worked examples.

Carbon index method The aim of this method is to provide for greater flexibility in the design of a dwelling, but with similar energy performance to that of the elemental method. Worked examples are shown in Appendices G and M to Approved Document L, based on calculations of CO_2 emissions in kg or tonnes per year relative to dwelling floor area. The formulae used are:

CF $= CO_2 \div (TFA + 45)$
CI $= 17.7 - (9 \log. CF)$

where: CF = Carbon Factor
CO_2 = Carbon dioxide emission in kg/year
TFA = Total floor area of dwelling (m^2)
CI = Carbon Index
log. = logarithm to the base of ten

The carbon index is expressed on a scale of 0 to 10, rounded to one deimal place. A high rating denotes high efficiency and all new dwellings should be rated at least 8.0. SAP worksheets in the Approved Document can be used to assess annual CO_2 emissions.

INSULATING MATERIALS

When selecting or specifying thermal insulation materials the following must be taken into consideration:

1. Thermal resistance of the material.
2. Need for a vapour control layer since insulating materials which become damp or wet, generally due to condensation, rapidly loose their insulation properties; therefore if condensation is likely to occur a suitable vapour control layer should be included in the detail. Vapour control layers should always be located on the warm side of the construction.
3. Availability of material chosen.
4. Ease of fixing or including the material in the general construction.
5. Appearance if visible.
6. Cost in relation to the end result and ultimate savings on fuel and/or heating installation.
7. Fire risk – all wall and ceiling surfaces must comply with the requirements of Building Regulation B2 – restriction of spread of flame over surfaces of walls and ceilings.

Insulating materials are made from a wide variety of materials and are available in a number of forms:

Insulating concrete   Basically a concrete of low density containing a large number of voids. This can be achieved by using lightweight aggregates such as clinker, foamed slag, expanded clay, sintered pulverized fuel ash, exfoliated vermiculite and expanded perlite, or alternatively an aerated concrete made by the introduction of air or gas into the mix. No fines concrete made by using lightweight or gravel aggregates between 20 and 10 mm size and omitting the fine aggregate is suitable for load bearing walls. Generally lightweight insulating concrete is used in the form of an *in-situ* screed to a structural roof or as lightweight concrete blocks for walls.

Loose fills   Materials which can be easily poured from a bag and levelled off between the joists with a shaped template. Materials include exfoliated vermiculite, fine glass fibrewool, mineral wool and cork granules. The depth required to satisfy current legislation is usually well above the top level of ceiling joists, therefore this type of insulation is now better used as a vertical fill between stud framing to dormer windows and walls in loft conversions. Most loose fills are rot and vermin proof as well as being classed as non-combustible.

Boards   Used mainly as dry linings to walls and ceilings either for self finish or direct decoration. Types include metallized polyester lined plasterboard, woodwool slabs, expanded polystyrene boards, thermal backed (expanded polystyrene/extruded polystyrene) plasterboard and fibreboards. Insulating fibreboards should be conditioned on site before fixing to prevent buckling and distortion after fixing. A suitable method is to expose the boards on all sides so that the air has a free passage around the sheets for at least 24 hours before fixing. During this conditioning period the boards must not be allowed to become wet or damp.

Quilts   Made from glass fibre or mineral wool bonded or stitched between outer paper coverings for easy handling. The quilts are supplied in rolls from 6.000 to 13.000 m long and cut to suit standard joist spacings. They are laid between and over the ceiling joists and can be obtained in two thicknesses, 100 mm for general use and 150 mm for use in roof spaces. They can be placed in two layers, the lower layer between the joists with another superimposed at right angles to the joists.

Insulating plasters   Factory produced premixed plasters which have lightweight perlite and vermiculite expanded minerals as aggregates, and require only the addition of clean water before application. They are only one-third the weight of sanded plasters, have three times the thermal insulation value and are highly resistant to fire. However, they can only be considered as a supplement to other insulation products as they have insufficient thickness to provide full insulation.

Foamed cavity fill   A method of improving the thermal insulation properties of an external cavity wall by filling the cavity wall with urea–formaldehyde resin foamed on site. The foam is formed using special apparatus by combining urea–formaldehyde resin, a hardener, a foaming agent and warm water. Careful control with the mixing and application is of paramount importance if a successful result is to be achieved; specialist contractors are normally employed. The foam can be introduced into the cavity by means of 25 mm bore holes spaced 1.000 m apart in all directions or by direct introduction into the open end of the cavity. The foam is a white cellular material containing approximately 99% by volume of air with open cells. The foam is considered to be impermeable and therefore unless fissures or cracks have occurred during the application it will not constitute a bridge of the cavity in the practical sense. Non-combustible water-repellent glass or rock wool fibres are alternative cavity fill materials. Manufacturer's approved installers use compressors to blow the fibres into new or existing cavity walls where both inner and outer leaves are constructed of masonry. The application of cavity fill must comply with the requirements of Building Regulation D1.

The most effective method of improving thermal comfort conditions within a building is to ensure that the inside surface is at a reasonably high temperature and this is best achieved by fixing insulating materials in this position.

Thermal insulation for buildings other than dwellings are also contained in the Building Regulations under Part L2. These are considered in *Advanced Construction Technology*.

THERMAL BRIDGING

Consistency of construction in the external elements is desirable to achieve a uniform heat transfer. It is, nevertheless, impractical to build in uniform layers and the routine is interrupted with 'bridges' between air voids and across solids such as timber in framed walls. This results in variations in thermal transmittance values (see proportional area calculation – previous chapter) which should be avoided as much as possible. Examples include metal or dense concrete lintels and gaps in the continuity of cavity insulation. Where this does occur, the thermal or cold bridge will incur a cooler surface temperature than the remainder of the wall, greater heat loss at this point and a possible dew point temperature manifesting in water droplets or condensation. The moisture will attract dust and dirt to contrast with adjacent clean areas. This is known as *pattern staining*.

The Approved Document to Part L of the Building Regulations makes specific reference to thermal bridging, indicating the areas most exposed. Further reference to the Building Regulations support document, *Thermal insulation: avoiding risks – 2nd edition*, provides numerous examples of acceptable forms of construction at vulnerable places.

Apart from examples of bad practice, such as using bricks to make up coursing in lightweight insulating concrete block inner leaf walls, the most common areas for thermal bridging are around door and window openings, junctions between ground floor and wall, and the junction between wall and roof. These are shown in Fig. 8.3.1.

As a measure of the significance of thermal bridging, the following formula can be applied:

$$\frac{0.3 \times \text{total length of opening surrounds}}{\text{total exposed surface areas}} + \text{average } U \text{ value}$$

[*Note*: Worked examples for calculating the average U value can be found in Appendix F of Approved Document L to the Building Regulations.]

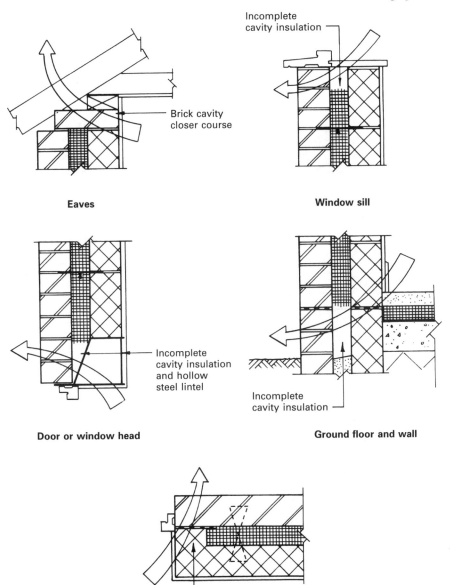

Figure 8.3.1 Thermal bridging – areas for concern

Figure 8.3.2   Thermal bridging – treatment at openings and junctions

If the calculation is less than the 'target *U* value' (see previous chapter) then thermal bridging is not significant.

Attention to detail at construction interfaces is the key to preventing thermal bridging. The details shown in Fig. 8.3.2 incorporate the necessary features to promote good practice.

DRAUGHT PROOFING

Draught proofing is required to prevent the infiltration of cold external air through leakage in the building envelope. It is also to limit heat losses through breaks in the continuity of construction.

The areas most exposed occur at the junction of different components in the same element, e.g. door and window abutments with walls, access hatches into the roof space or eaves cupboards, etc. The fit between window sashes and frames, and doors and frames, are also very vulnerable to air leakage. The closeness of fit should be the subject of quality control at the point of manufacture, but this alone is insufficient to satisfy Part L of the Building Regulations. All opening units should have purpose-made seals or draught excluders.

Particular attention must be provided at the points where service pipes and ducts penetrate the structure. Gaps must be made good with appropriate filling and a flexible sealant applied between pipe or duct and adjacent filling. Dry linings, too, must receive special attention. It is not sufficient to secure these to background walls with only plaster dabs. Continuous plaster seals are required at abutments with walls, floors, ceilings and openings for services, windows and doors. Some areas and methods of treatment are shown in Fig. 8.4.1.

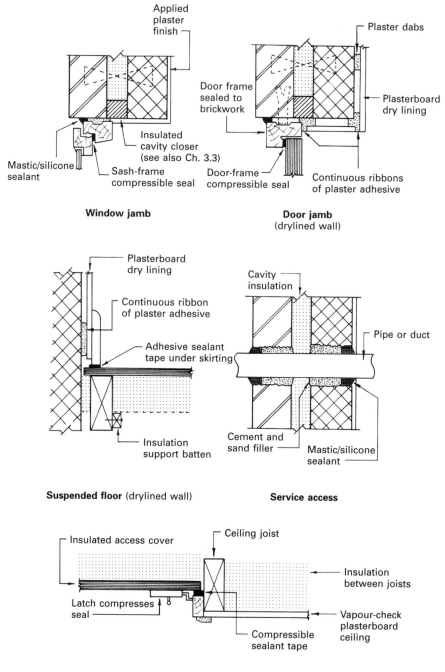

Window jamb

Door jamb
(drylined wall)

Suspended floor (drylined wall)

Service access

Loft hatch

Figure 8.4.1   Prevention of air infiltration

PART 9

FRAMED BUILDINGS

SIMPLE FRAMED BUILDINGS

The purpose of any framed building is to transfer the loads of the structure plus any imposed loads through the members of the frame to a suitable foundation. This form of construction can be clad externally with lightweight non-loadbearing walls to provide the necessary protection from the elements and to give the required degree of comfort in terms of sound and thermal insulation. Framed buildings are particularly suitable for medium- and high-rise structures and for industrialized low-rise buildings such as single-storey factory buildings.

Frames can be considered under three headings:

- Plane frames.
- Space frames.
- Skeleton frames.

PLANE FRAMES

These are fabricated in a flat plane and are usually called trusses or girders according to their elevation shape. They are designed as a series of connected rigid triangles which gives a lightweight structural member using the minimum amount of material; main uses are in roof construction and long span beams of light loading.

SPACE FRAMES

These are similar in conception to a plane frame but are designed to span in two directions as opposed to the one-direction spanning of the plane frame. A variation of the space frame is the space deck which consists of a series of linked pyramid frames forming a lightweight roof structure. For details of these forms of framing, refer to *Advanced Construction Technology*.

SKELETON FRAMES

Basically, these are a series of rectangular frames placed at right angles to one another so that the loads are transmitted from member to member until they are transferred through the foundations to the subsoil. Skeleton frames can be economically constructed of concrete or steel or a combination of the two. Timber skeleton frames, although possible, are generally considered to be uneconomic in this form. The choice of material for a framed structure can be the result of a number of factors such as site conditions, economics, availability of labour and materials, time factor, statutory regulations, capital costs, maintenance costs and personal preference.

FUNCTIONS OF SKELETON FRAME MEMBERS

- **Main beams**   Span between columns and transfer the live and imposed loads placed upon them to the columns.
- **Secondary beams**   Span between and transfer their loadings to the main beams. Primary function is to reduce the spans of the floors or roof being supported by the frame.
- **Tie beams**   Internal beams spanning between columns at right angles to the direction of the main beams and having the same function as a main beam.
- **Edge beams**   As tie beam but spanning between external columns.
- **Columns**   Vertical members which carry the loads transferred by the beams to the foundations.
- **Foundation**   The base(s) to which the columns are connected and serve to transfer the loadings to a suitable loadbearing subsoil.
- **Floors**   May or may not be an integral part of the frame; they provide the platform on which equipment can be placed and on which people can circulate. Besides transmitting these live loads to the supporting beams they may also be required to provide a specific fire resistance, together with a degree of sound and thermal insulation.
- **Roof**   Similar to floors but its main function is to provide a weather-resistant covering to the uppermost floor.
- **Walls**   The envelope of the structure which provides the resistance to the elements, entry of daylight, natural ventilation, fire resistance, thermal insulation and sound insulation.

The three major materials used in the construction of skeleton frames – namely, reinforced concrete, precast concrete and structural steel work – are considered in detail in the following chapters.

REINFORCED CONCRETE FRAMES

CONCRETE

Plain concrete is a mixture of cement, fine aggregate, coarse aggregate and water. Concrete sets to a rock-like mass due to a chemical reaction which takes place between the cement and water, resulting in a paste or matrix which binds the other constituents together. Concrete gradually increases its strength during the curing or hardening period to obtain its working strength in about twenty-eight days if ordinary Portland cement is used. The strength achieved will depend on a number of factors such as:

Type of cement used   In all cases the cement should conform to the appropriate British Standard.

Type and size of aggregates   In general, aggregates should comply with BS 882 for quality and grading. BS 812 provides tests for aggregates and defines nominal maximum sizes into categories of 40, 20, 14 and 10 mm. 5 mm is also included, although used mainly for precast concrete products.

Water   This should be clean, free from harmful matter and comply with the requirements of BS 3148.

Use of admixtures   Individually, admixtures can be used as accelerators, for air entrainment to give a weight reduction and added protection against water penetration, chemical and fungal attack. The instructions given by the manufacturer or engineer must be carefully followed since many admixtures if incorrectly used can have serious adverse effects on the hardened concrete.

Water/cement ratio   A certain amount of water is required to hydrate the cement and any extra water is needed only to produce workability. The workability of fresh concrete should be such that it can be handled, placed and compacted so that it will surround any reinforcement and completely fill the formwork. Workability can be specified in terms of a slump test, compacting factor test or VB consistometer test and should not vary beyond the values for the maximum free water/cement ratio given in BS 5328: Part 1.

Grade strength   Concrete should be specified as designated, designed or prescribed mix, with the grade required in accordance with BS 5328: Part 2 (see also Chapter 2.4 of this book). Grades are numerically related to characteristic strength or cube test strength taken at 28 days. This is for concrete with any type of cement except high-alumina cement which hardens rapidly and has structural limitations – see Chapter 2.4. The grades and general application include:

- Grades 7 and 10 for plain unreinforced concrete. C10 is the minimum grade normally acceptable, provided the cement content is at least 175 kg/m^3 for designated mixes and 210 kg/m^3 for other concrete specifications.
- Grades 15 and 20 for reinforced concrete with lightweight aggregates.
- Grade 25 for reinforced concrete with dense aggregates.
- Grades 30 to 50 for prestressed concrete.

The characteristic strength of grade 7 concrete is 7.0 N/mm^2, and for grade 10 concrete it is 10.0 N/mm^2; similar values can be deduced for the other grades listed above.

Plain concrete in common with other brittle materials has a greater crushing or compressive strength than tensile strength. The actual ratio varies but plain concrete is generally considered to be ten times stronger in compression than in tension. If a plain concrete member is loaded so that tension is induced it will fail in tension when the compressive strength of concrete has only reached one-tenth of its ultimate value. If this weakness in tension can be reinforced in such a manner that the tensile resistance is raised to a similar value as its compressive strength the member will be able to support a load ten times that of plain concrete or alternatively for any given load a smaller section can be used if concrete is reinforced.

REINFORCEMENT

Any material specified for use as a reinforcement to concrete must fulfil certain requirements if an economic structural member is to be constructed. These basic requirements are:

1. It must have tensile strength.
2. It must be capable of achieving this tensile strength without undue strain.
3. It must be of a material that can be easily bent to any required shape.
4. Its surface must be capable of developing an adequate bond between the concrete and the reinforcement to ensure that the required design tensile strength is obtained.
5. A similar coefficient of thermal expansion is required to prevent unwanted stresses being developed within the member due to temperature changes.
6. It must be available at a reasonable cost which must be acceptable to the overall design concept.

The material which meets all the above requirements is steel in the form of bars, and is supplied in two basic types, namely mild steel and high-yield steel. Hot-rolled steel bars are covered by BS 4449 which specifies a characteristic strength of 250 N/mm^2 for mild steel and 460 N/mm^2 for high-yield steel. The surface of mild steel provides adequate bond but the bond of high-yield bars, being more critical with the higher stresses developed, is generally increased by rolling onto the surface of the bar longitudinal or transverse ribs. Grade 460A can be identified by having at least two rows of parallel transverse ribs, while grade 460B is similar but has at least one row of ribs contrary to the others. As an alternative to hot-rolled steel bars, cold-worked steel bars can be used. When bars are cold worked they become harder, stiffer and develop a higher tensile strength.

The range of nominal sizes available for both round and deformed bars are 6, 8, 10, 12, 16, 20, 25, 32 and 40 mm with length specified by the purchaser. For pricing purposes the 16 and 20 mm bars are taken as basic with the diameters on either side becoming more expensive as the size increases or decreases. A good design will limit the range of diameters used together with the type of steel chosen to achieve an economic structure and to ease the site processes of handling, storage, buying and general confusion that can arise when the contractor is faced with a wide variety of similar materials.

The bending of reinforcement can be carried out on site by using a bending machine which shapes the cold bars by pulling them round a mandrel. Small diameters can also be bent round simple jigs such as a board with dowels fixed to give the required profile; large diameters may need a power-assisted bending machine. Bars can also be supplied ready bent and labelled so that only the fabrication processes takes place on site.

The bent reinforcement should be fabricated into cages for columns and beams and into mats for slabs and walls. Where the bars cross or intersect one another they should be tied with soft iron wire, fixed with special wire clips or tack welded to maintain their relative positions. Structural members which require only small areas of reinforcement can be reinforced with steel fabric which can be supplied in sheets or rolls.

Steel fabric for the reinforcement of concrete is covered by BS 4483, which gives four basic preferred types: 'A' for square mesh (floor slabs), 'B' for rectangular structural mesh (also for slabs), 'C' for long mesh (roads and pavements) and 'D' a small-diameter wrapping mesh (binding concrete fire proofing). The fabric is factory made by welding or interweaving wires complying with the requirements of BS 4482 to form sheets with a length of 4.800 m and a width of 2.400 m or alternatively rolls of 48.000 and 72.000 m in length with a common width of 2.400 m. Each type has a letter prefix which is followed by a reference number which is the total cross sectional area of main bars in mm^2 per metre width. Typical examples of reinforcing bars and fabric are shown in Fig. 9.2.1.

Before placing reinforcement into the formwork it should be brushed free of all loose rust and mill scale and be free of grease or oil, as the presence of any of these on the surface could reduce the bond and hence the strength of the reinforced concrete. Reinforcement must have a minimum cover of concrete to give the steel

Square twisted bar

Plain round bar

Ribbed bar

Twisted ribbed bar

200

200

10, 8, 7, 6 & 5 mm
dia. – main and cross wires similar

Square mesh fabric – BS prefix 'A'

200

100

Cross wires –
8 & 7 mm dia.

Main wires – 12, 10, 8,
7, 6 & 5 mm dia.

Structural mesh fabric – BS prefix 'B'

400

100

Cross wires –
6 & 5 mm dia.

Main wires – 10, 9, 8,
7 & 6 mm dia.

Long mesh fabric – BS prefix 'C'

Figure 9.2.1   Typical reinforcing bars and welded fabric

protection from corrosion due to contact with moisture and to give the structural
member a certain degree of fire resistance. Nominal cover to reinforcement should
always be equal to the size of the bar being used or where groups of bars are used
at least the size of the largest diameter. BS 8110, Tables 3.3 and 3.4, set out the
recommended nominal amount of cover, in relationship to the exposure/durability
and periods of fire-resistance criteria, respectively, for various concrete grades.
The Building Regulations, Approved Document B, establishes the minimum fire
resistance for various purpose groups of buildings. It must be noted that the

dimensions of structural members are also of importance so that failure of the concrete due to the high temperatures encountered during a fire is avoided before the reinforcement reaches its critical temperature. Table A3 of Approved Document B and Table 3.5 and Fig. 3.2 in BS 8110: Part 1 give suitable dimensions for various fire-resistance periods and methods of construction of reinforced concrete members.

To maintain the right amount of concrete cover during construction small blocks of concrete may be placed between the reinforcement and the formwork; alternatively, plastic clips or spacers rings can be used. Where top reinforcement has to be retained in position (such as in a slab) cradles or chairs made from reinforcing bar may have to be used. All forms of spacers must be of a material which will not lead to corrosion of the reinforcement or cause spalling of the hardened concrete.

DESIGN

The design of reinforced concrete is the prerogative of the structural engineer and is a subject for special study but the technologist should have an understanding of the principles involved. The designer can by assessing the possible dead and live loads on a structural member calculate the reactions and effects such loadings will have on the member. This will indicate areas of tensile stress and critical locations for reinforcement. Calculations are based on the recommendations and design tables in BS 8110: Parts 1, 2 and 3: *Structural use of concrete*, together with established formulae for determination of bending moments, shear forces and the area of steel required. Typical examples of these forces for simple situations are shown in Fig. 9.2.2.

REINFORCEMENT SCHEDULES AND DETAILS

Once the engineer has determined the reinforcement required, detail drawings can be prepared to give the contractor the information required to construct the structure. The drawings should give the following information:

1. Sufficient cross-reference to identify the member in relationship to the whole structure.
2. All the necessary dimensions for design and fabrication of formwork.
3. Details of the reinforcement.
4. Minimum cover of concrete over reinforcement.
5. Concrete grade required if not already covered in the specification.

Reinforced concrete details should be prepared so that there is a distinct definition between the lines representing the outline of the member and those representing the reinforcement. Bars of a common diameter and shape are normally grouped together with the same reference number when included in the same member. To simplify the reading of reinforced concrete details it is common

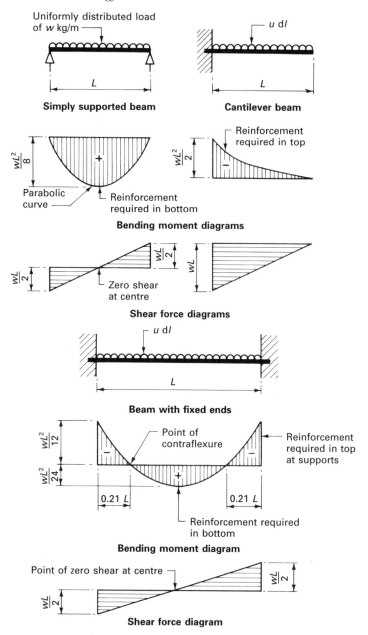

Figure 9.2.2   Bending moment and shear force diagrams

practice to show only one bar of each group in full together with the last bar position (see Fig. 9.2.3).

The bars are normally bent and scheduled in accordance with the recommendations of BS 4466 which gives details of the common bending shapes,

Elevation – beam 1-3 No. thus

Note: Cover to main bars 25 mm

Member	Bar mark	Type & size	No. of mbrs	No. in each	Total no.	Length of each bar[†]	Shape. All dimensions[*] are in accordance with BS 4466
Beam 1	1	R20	3	2	6	2.660	⌐2.300⌐
	2	R16	3	1	3	1.400	Straight
	3	R10	3	2	6	2.300	Straight
	4	R10	3	16	48	1.000	250 ☐ 150
[†] Specified to nearest 25 mm						[*] Specified to nearest 5 mm	

Figure 9.2.3 Typical reinforced concrete beam details and schedule

the method of setting out the bending dimensions, the method of calculating the total length of the bar required together with a shape code for use with data-processing routines. A preferred form of bar schedule is also given, which has been designed to give easy cross-reference to the detail drawing.

Reinforcement on detail drawings is annotated by a coding system to simplify preparation and reading of the details, for example:

9 R 1201–300 which can be translated as:

9 = total number of bars in the group
12 = diameter in mm
01 = bar mark number
300 = spacing centre to centre
R = mild steel round bar (250 N/mm²)

The code letter R could be replaced by code letter T which is used for high-yield round bars and high-yield square twisted bars; other types of bars are coded S for stainless steel and W for wire reinforcement (4 to 12 mm dia.). A typical reinforced concrete beam detail and schedule is shown in Fig. 9.2.3. All other reinforced concrete details shown in this volume are intended to show patterns of reinforcement rather than detailing practice and are therefore shown in full without reference to bar diameters and types.

HOOKS, BENDS AND LAPS

To prevent bond failure, bars should be extended beyond the section where there is no stress in the bar. The length of bar required will depend upon such factors as grade of concrete, whether the bar is in tension or compression and if the bar is deformed or plain. Hooks and bends can be used to reduce this anchorage length at the ends of bars and should be formed in accordance with the recommendations of BS 4466 (see Fig. 9.2.4).

Where a transfer of stress is required at the end of a bar the bars may be welded or lapped.

BS 8110 recommends that laps and joints should only be made by the methods specified in the contract and at the positions shown on the drawings and as agreed by the engineer.

REINFORCED CONCRETE BEAMS

Beams can vary in their complexity of design and reinforcement from the very simple beam formed over an isolated opening such as those shown in Figs 9.2.3 and 9.2.5 to the more common form encountered in frames where the beams transfer their loadings to the columns (see Fig. 9.2.6).

When tension is induced into a beam the fibres will lengthen until the ultimate tensile strength is reached, when cracking and subsequent failure will occur. With a uniformly distributed load the position and value of tensile stress can easily be calculated by the structural engineer, but the problem becomes more complex when heavy point loads are encountered.

The correct design of a reinforced concrete beam will ensure that it has sufficient strength to resist both the compression and tensile forces encountered in the outer fibres, but it can still fail in the 'web' connecting the compression and tension

h = 9d min. to
nearest 10 mm over
min. = 100

Mild steel bar

n = 5d min. to
nearest 10 mm over
min. = 100

Mild steel bar

h = 11d min. to
nearest 10 mm over
min. = 100

High-yield bar

n = 5.5d min. to
nearest 10 mm over
min. = 100

High-yield bar

Figure 9.2.4 Standard hooks and bends

areas. This form of failure is called shear failure and is in fact diagonal tension. Concrete has a limited amount of resistance to shear failure and if this is exceeded reinforcement must be added to provide extra resistance. Shear occurs at or near the supports as a diagonal failure line at an angle of approximately 45° to the

Simple beam with shear reinforcement

Nominal diameter anchor bars – Mk 3

Stirrups at close centres to assist bent up shear bar – Mk 4

Stirrups at nominal centres Mk 4

Mk 2 bent up as shear bar

Mk 1

Bearing

Main bars – Mks 1 & 2

Clear span

Bearing

Section mid-span

Width

Section near support

Width

Depth

Cantilever beam to canopy

Main bars cranked down to avoid edge beam top steel

Stirrups at nominal centres – Mk 3

Main bars – Mk 2

Nominal diameter anchor bars – Mk 1

RC edge beam

Clear span

Bearing

Typical section

Width

RC slab

Depth varies

222

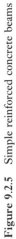

Figure 9.2.5 Simple reinforced concrete beams

Figure 9.2.6 Reinforced concrete beam with heavy reinforcement

Binders Mk 5 at nominal centres

Binders Mk 5 at close centres

Mk 2

RC edge beam

RC column

Mk 3

Nominal diameter anchor bars Mk 4 cranked under secondary beam steel

Mk 1

Shear bar Mk 3

Spacer bars

Elevation

Binders Mk 5 at close centres

RC slab

Mk 3

Main bars Mks 1 & 2

Mk 1

Mk 2

RC edge beam

RC column

RC slab

4 4

2 2

5

1 1

RC column

RC edge beam

Section at support

RC slab

RC secondary beam

4 4

5

3

1 2 2 1

Spacer bar diameter not less than main bar diameter or aggregate size

Section at mid-span

horizontal and sloping downwards towards the support. A useful fact to remember is that zero shear occurs at the point of maximum bending (see Fig. 9.2.2).

Reinforcement to resist shearing force may be either stirrups or inclined bars, or both. The total shearing resistance is the sum of the shearing resistances of the inclined bars and the stirrups, calculated separately if both are provided. Inclined or bent up bars should be at 45° to the horizontal and positioned to cut the anticipated shear failure plane at right angles. These may be separate bars or alternatively main bars from the bottom of the beam which are no longer required to resist tension which can be bent up and carried over or onto the support to provide the shear resistance (see Figs 9.2.5 and 9.2.6). Stirrups or binders are provided in beams, even where not required for shear resistance, to minimize shrinkage cracking and to form a cage for easy handling. The nominal spacing for stirrups must be such that the spacing dimension used is not greater than the lever arm of the section, which is the depth of the beam from the centre of the compression area to the centre of the tension area or 0.75 times the effective depth of the beam, which is measured from the top of the beam to the centre of the tension reinforcement. If stirrups are spaced at a greater distance than the lever arm it would be possible for a shearing plane to occur between consecutive stirrups, but if the centres of the stirrups are reduced locally about the position at which shear is likely to occur several stirrups may cut the shear plane and therefore the total area of steel crossing the shear plane is increased to offer the tensile resistance to the shearing force (see Figs 9.2.5 and 9.2.6).

REINFORCED CONCRETE COLUMNS

A column is a vertical member carrying the beam and floor loadings to the foundation and is a compression member. Since concrete is strong in compression it may be concluded that provided the compressive strength of the concrete is not exceeded no reinforcement will be required. For this condition to be true the following conditions must exist:

1. Loading must be axial.
2. Column must be short, which can be defined as a column where the ratio of its effective height to its thickness does not exceed 12 (effective height may be calculated by formula and tables in Section 3.8.1.6 of BS 8110: Part 1).
3. Cross-section of the column must be large.

These conditions rarely occur in framed buildings, consequently bending is induced and the need for reinforcement to provide tensile strength is apparent. Bending in columns may be induced by one or more of the following conditions:

1. Load coupled with the slenderness of the column; a column is considered to be slender if the ratio of effective height to thickness exceeds 12.
2. Reaction to beams upon the columns; as the beam deflects it tends to pull the column towards itself thus inducing tension in the far face.
3. The reaction of the frame to wind loadings both positive and negative.

The minimum number of main bars in a column should not be less than four for rectangular columns and six for circular columns with a total cross-sectional area of not less than 0.8% of the cross-sectional area of the column and a minimum diameter of 12 mm. To prevent the slender main bars from buckling, and hence causing spalling of the concrete, links or binders are used as a restraint. These should be at least 6 mm diameter and not less than one quarter of the largest main bar diameter. The pitch or spacing is no greater than the least of:

- the least lateral column dimension
- 12 times the smallest longitudinal reinforcement
- 300 mm

All bars in compression should be tied by a link passing around the bar in such a way that it tends to move the bar towards the centre of the column; typical arrangements are shown in Fig. 9.2.7.

Where the junction between beams and columns occurs there could be a clash of steel since bars from the beam may well be in the same plane as bars in the columns. To avoid this situation one group of bars must be offset or cranked into another plane; it is generally considered that the best practical solution is to crank the column bars to avoid the beam steel; typical examples of this situation together with a method using straight bars are shown in Fig. 9.2.8. A similar situation can occur where beams of similar depth intersect; see cantilever beam example in Fig. 9.2.5.

REINFORCED CONCRETE SLABS

A reinforced concrete slab will behave in exactly the same manner as a reinforced concrete beam and it is therefore designed in the same manner. The designer will analyse the loadings, bending moments, shear forces and reinforcement requirements on a slab strip 1.000 m wide. In practice the reinforcement will be fabricated to form a continuous mat. For light loadings a mat of welded fabric could be used.

There are three basic forms of reinforced concrete slabs, namely:

- Flat slab floors or roofs.
- Beam and slab floors or roofs.
- Ribbed floors or roofs – see Chapter 4.4.

Flat slabs

These are basically slabs contained between two plain surfaces and can be either simple or complex. The design of the complex form is based upon the slab acting as a plate in which the slab is divided into middle and column strips; the reinforcement being concentrated in the latter strips.

Simple flat slabs can be thick and heavy but have the advantage of giving clear ceiling heights since there are no internal beams. They are generally economic up to

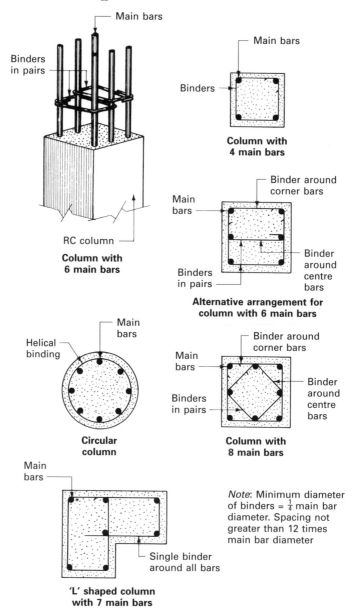

Figure 9.2.7 Typical reinforced concrete column binding arrangements

spans of approximately 9.000 m and can be designed to span one way, that is across the shortest span, or to span in two directions. These simple slabs are generally designed to be simply supported, that is, there is no theoretical restraint at the edges and therefore tension is not induced and reinforcement is not required. However, it is common practice to provide some top reinforcement at the supports

Figure 9.2.8 Reinforced concrete column and beam junctions

as anti-crack steel should there, in practice, be a small degree of restraint. Generally this steel is 50% of the main steel requirement and extends into the slab for 0.2 m of the span. An economic method is to crank up 50% of the main steel or every alternate bar over the support since the bending moment would have reduced to such a degree at this point it is no longer required in the bottom of the slab. If

Binders from beam bent into top of slab for $\frac{1}{6}$ to $\frac{1}{4}$ of span

Distribution bars

Beam over opening

Beam main bars

Distribution bars

Main bars cranked at one end with alternate bars reversed – detail similar to right hand support

Distribution bars – alternate bars cranked into top of slab for $\frac{1}{6}$ to $\frac{1}{4}$ of span

Typical RC flat slab

Mk 1 →

Beam reinforcement

Distribution bars Mk 2

Distribution bars Mk 2

Main bars Mk 1 – cranked at one end with alternate bars reversed

Mk 3 →

Beam binders

Beam main bars

RC kerb

Main bars Mk 3

← Mk 1

Distribution bars

Edge trimming bar Mk 4

Typical RC beam and slab with cantilever

Figure 9.2.9 Typical reinforced concrete slab details

there is an edge beam the top steel can also be provided by extending the beam binders into the slab (see Fig. 9.2.9).

Beam and slab

By adopting this method of design large spans are possible and the reinforcement is generally uncomplicated. A negative moment will occur over the internal supports necessitating top reinforcement; as with the flat slabs, this can be provided by cranked bars (see Fig. 9.2.9). Each bar is in fact cranked but alternate bars are reversed thus simplifying bending and identification of the bars. Alternatively a separate mat of reinforcement supported on chairs can be used over the supports.

FORMWORK

Formworks for *in-situ* concrete work may be described as a mould or box into which wet concrete can be poured and compacted so that it will flow and finally set to the inner profile of the box or mould. It is important to remember that the inner profile must be opposite to that required for the finished concrete so if, for example, a chamfer is required on the edge a triangle fillet must be inserted into the formwork.

To be successful in its function formwork must fulfil the following requirements:

1. It should be strong enough to support the load of wet concrete which is generally considered to be approximately 2,400 kg/m^3.
2. It must not be able to deflect under load which would include the loading of wet concrete, self-weight and any superimposed loads such as operatives and barrow runs over the formwork.
3. It must be accurately set out; concrete being a fluid when placed, it will take up the shape of the formwork which must therefore be of the correct shape, size and in the right position.
4. It must have grout-tight joints. Grout leakage can cause honeycombing of the surface or produce fins which have to be removed. The making good of defective concrete surfaces is both time consuming and costly. Grout leakage can be prevented by using sheet materials and sealing the joints with flexible foamed polyurethane strip or by using a special self-adhesive tape.
5. Form sizes should be designed so that they are the maximum size which can easily be handled by hand or by a mechanical lifting device.
6. Material must be chosen so that it can be easily fixed using either double-headed nails, round wire nails or wood screws. The common method is to use nails and these should be at least two-and-a-half times the thickness of the timber being nailed, in length.
7. The design of the formwork units should be such that they can easily be assembled and dismantled without any members being trapped.

MATERIALS

The requirements for formwork enumerated above make timber the most suitable material for general formwork. It can be of board form either wrot or unwrot, depending on whether a smooth or rough surface is required.

Softwood boards used to form panels for beam and column sides should be joined together by cross-members over their backs at centres not exceeding twenty-four times the board's thickness.

The moisture content of the timber should be between 15 and 20% so that the moisture movement of the formwork is reduced to a minimum. If the timber is dry it will absorb moisture from the wet concrete which could weaken the resultant concrete member. It will also cause the formwork to swell and bulge which could give an unwanted profile to the finished concrete. If timber with a high moisture content is used it will shrink and cup, which could result in open joints and a leakage of grout.

Plywood is extensively used to construct formwork units since it is strong, light and supplied in sheets of 1.200 m wide with standard lengths of 2.400, 2.700 and 3.000 m. The quality selected should be an exterior grade and the thickness related to the anticipated pressures so that the minimum number of strengthening cleats on the back are required.

Chipboard can also be used as a formwork material but because of its lower strength will require more supports and stiffeners. The number of uses which can be obtained from chipboard forms is generally less than plywood, softwood boarding or steel.

Steel forms are generally based upon a manufacturer's patent system and within the constraints of that system are an excellent material. Steel is not as adaptable as timber but if treated with care will give thirty or forty uses, which is approximately double that of similar timber forms.

MOULD OILS AND EMULSIONS

Two defects which can occur on the surface of finished concrete are:

1. **Blow holes**   These are small holes, of less than 15 mm in diameter, caused by air being trapped between the formwork and the concrete face.
2. **Uneven colour**   This is caused by irregular absorption of water from the wet concrete by the formwork material. A mixture of old and new material very often accentuates this particular defect.

Mould oils can be applied to the inside surface of the formwork to alleviate these defects. To achieve a uniform colour an impervious material or lining is recommended, but this will increase the risk of blow holes. Mould oils are designed to overcome this problem when using steel forms or linings by encouraging the trapped air to slide up the face of the formwork. A neat oil will encourage blow holes but will discourage uneven colour, whereas a mould oil incorporating an emulsifying agent will discourage blow holes and reduce uneven colouring. Great care must be taken when applying some mould oils since over-oiling may cause some retardation of the setting of the cement. Emulsions are either drops of water in oil or, conversely, drops of oil in water and are easy to apply but should not be used in conjunction with steel forms since they encourage rusting. It should be noted that, generally, mould oils and emulsions also act as release agents and therefore it is essential that the oil or emulsion is only applied to the formwork and not to the reinforcement since this may cause a reduction of bond.

FORMWORK LININGS

To obtain smooth, patterned or textured surfaces the inside of a form can be lined with various materials such as oil-tempered hardboard, moulded rubber, moulded PVC and glass-fibre-reinforced polyester; the latter is also available as a complete form mould. When using any form of lining the manufacturer's instructions regarding sealing, fixing and the use of mould oils must be strictly followed to achieve a satisfactory result.

TYPES OF FORMWORK

FOUNDATION FORMWORK

Foundations to a framed building consist generally of a series of isolated bases or pads, although if these pads are close together it may be more practicable to merge them together to form a strip. If the subsoil is firm and hard it may be possible to excavate the trench or pit for the foundations to the size and depth required and cast the concrete against the excavated faces. Where this method is not practicable formwork will be required. Side and end panels will be required and these should be firmly strutted against the excavation faces to resist the horizontal pressures of the wet concrete and to retain the formwork in the correct position. Ties will be required across the top of the form as a top restraint and these can be utilized to form the kicker for a reinforced concrete column or as a template for casting in the holding-down bolts for precast concrete or structural steel columns (see Fig. 9.3.1).

COLUMN FORMWORK

A column form or box consists of a vertical mould which has to resist considerable horizontal pressures in the early stages of casting. The column box should be located against a 75 mm-high plinth or kicker which has been cast monolithic with the base or floor. The kicker not only accurately positions the formwork but also prevents the loss of grout from the bottom edge of the form. The panels forming the column sides can be strengthened by using horizontal cleats or vertical studs which are sometimes called soldiers. The form can be constructed to the full storey

22 or 25 mm thick square
edge or t & g boarding

75 × 32 cleats at not more
than 24 times board
thickness centres

Typical boarded formwork panel

21 mm thick plywood or
blockboard face

50 × 50 studs
at 600 c/c

50 × 50 framing

Typical framed formwork panel

Battered excavation

Straight cut
excavation

Thrust
pad

75 × 32 ties as
required – could be
framed to form kicker
or used to position
holding-down bolts

Thrust
pad

Formwork to
edge faces

75 × 50 struts

Struts

Typical foundation formwork

Figure 9.3.1 Formwork to foundations

height of the column with cut-outs at the top to receive the incoming beam
forms. The thickness of the sides does not generally provide sufficient bearing
for the beam boxes and therefore the cut-outs have a margin piece fixed around the
opening to provide extra bearing (see Fig. 9.3.2). It is general practice however to

Opening for beam form
Fillet to form chamfer
Sides of sheet material
Margin piece
Soldiers or studs
Alternative panel format
Struts as required
Yokes or clamps positioned to resist pressures – see Fig. 9.3.3
Cleats – taken beyond width of panel side to form rebate
Formwork panels of boards or sheet material
Access pocket piece for cleaning out base of column form before pouring concrete
Foot of form located around kicker

Figure 9.3.2   Column formwork principles

cast the columns up to the underside of the lowest beam soffit and to complete the top of the column at the same time as the beam using make-up pieces to complete the column and receive the beam intersections. The main advantage of casting full height columns is the lateral restraint provided by the beam forms, the disadvantage being the complexity of the formwork involved.

Column form

100 × 75 yoke

Cleats

16 mm dia. threaded dowel

Hardwood wedges

Plate washer to both ends

100 × 75 yoke

Wedged oversize halving joint

Oversize halving joint

Bolted halving joint to opposite corners

Typical timber yokes

Security chain or wire

Column box or form

Steel wedge

32 × 8 slots

Range of clamp sizes: 300 min. to 1,400 max.

Slotted steel blade or arm

Typical column clamp

Figure 9.3.3 Typical column yokes and clamps

Column forms are held together with collars of timber or metal called *yokes* in the case of timber and *clamps* when made of metal. Timber yokes are purpose made whereas steel column clamps are adjustable within the limits of the blades (see Fig. 9.3.3).

The spacing of the yokes and clamps should vary with the anticipated pressures, the greatest pressure occurring at the base of the column box. The actual pressure will vary according to:

- Rate of placing.
- Type of mix being used – generally the richer the mix the greater the pressure.
- Method of placing – if vibrators are used pressures can increase up to 50% over hand placing and compacting.
- Air temperature – the lower the temperature the slower is the hydration process and consequently higher pressures are encountered.

Some preliminary raking strutting is required to plumb and align the column forms in all situations. Free-standing columns will need permanent strutting until the concrete has hardened but with tied columns the need for permanent strutting must be considered for each individual case.

Shaped columns will need special yoke arrangements unless they are being formed using a patent system. Typical examples of shaped column forms are shown in Fig. 9.3.4.

BEAM FORMWORK

A beam form consists of a three-sided box which is supported by cross-members called *headtrees* which are propped to the underside of the soffit board. In the case of framed buildings support to the beam box is also provided by the column form. The soffit board should be thicker than the beam sides since this member will carry the dead load until the beam has gained sufficient strength to be self-supporting. Soffit boards should be fixed inside the beam sides so that the latter can be removed at an early date; this will enable a flow of air to pass around the new concrete and speed up the hardening process and also release the formwork for re-use at the earliest possible time. Generally the beam form is also used to support the slab formwork and the two structural members are then cast together. The main advantage of this method is that only one concrete operation is involved, although the complexity of the formwork is increased. If the beams and slabs are carried out as separate operations there is the possibility of a shear plane developing between the beam and floor slab; it would be advisable to consult the engineer before adopting this method of construction. Typical examples of beam forms are shown in Fig. 9.3.5.

Structural steelwork has to be protected against corrosion and fire; one method is to encase the steel section with concrete. The steel frame is erected before casting the concrete encasement and, in the case of beams, it is possible to suspend the form box from the steel section by using a metal device called a hanger fixing or alternatively using a steel column clamp or timber yoke (see Fig. 9.3.6). The hanger fixings are left embedded in the concrete encasing but the bolts and plate washers are recoverable for re-use. If only a haunch is cast around the bottom flange then the projecting hanger fixing wires can be cut off level with the concrete haunch or the floor units can be slotted to receive them.

Octagonal column

Circular column

Rebated column

Figure 9.3.4 Shaped column forms and yokes

Simple beam or lintel formwork

Edge beam and slab formwork

Figure 9.3.5 Typical beam formwork

Hanger fixings at 600 c/c

Universal beam

Outline of slab

Support joists

Soffit form

Beam form

Hanger bolts with square plate washers

Plastic or precast concrete spacer blocks to provide required cover

Hanger wires welded to nut

Wire nut

Hanger fixings are made to suit BS 4 universal beam profiles – 12 mm should be allowed for gap between hanger fixing and beam soffit

Typical hanger fixing

Security chain and wedge

Universal beam

Haunch form

In-situ concrete haunch to provide protection to lower flange and seating for precast concrete floor units

Plastic or precast concrete spacer blocks

Steel column clamps at 600 c/c – alternative timber column yokes

Figure 9.3.6 Suspended formwork

Figure 9.3.7 Typical beam and slab formwork

Internal beam

75 × 32 strut

75 × 32 runner or stringer

21 mm thick plywood soffit

Beam or fork head

Inner tube

Pin

Collar   75 × 50 cleat

Alternative to timber props – adjustable steel props – 5 sizes with range of 1.800 to 4.870

150 × 150 base plate

Outline of slab

150 × 50 joists at 600 c/c

100 × 75 crosshead or headtree

75 × 32 brace

Outer tube

Bracing if required

225 × 75 sole plate

Edge beam

15 mm thick plywood beam sides

21 mm thick beam soffit

150 × 75 props at 1.200 c/c

Folding wedges

SLAB FORMWORK

Floor or roof slab formwork is sometimes called shuttering and consists of panels of size that can be easily handled. The panels can be framed or joisted and supported by the beam forms with any intermediate propping which is required (see Fig. 9.3.7). Adjustment for levelling purposes can be carried out by using small folding wedges between the joists or framing and the beam box.

SITEWORK

When the formwork has been fabricated and assembled the interior of the forms should be cleared of all rubbish, dirt and grease before the application of any mould oil or releasing agent. All joints and holes should be checked to ensure that they are grout tight.

The distance from the mixer to the formwork should be kept as short as possible to maintain the workability of the mix and to avoid double handling as far as practicable. Care must be taken when placing and compacting the concrete to ensure that the reinforcement is not displaced. The depth of concrete that can be placed in one lift will depend upon the mix and section size. If vibrators are used as the means of compaction this should be continuous during the placing of each batch of concrete until the air expulsion has ceased and care must be taken since over-vibrating concrete can cause segregation of the mix.

The striking or removal of formwork should only take place upon instruction from the engineer or agent. The appropriate time at which it is safe to remove formwork can be assessed by tests on cubes taken from a similar batch mixed at the time the concrete was poured and cured under similar conditions. The characteristic cube strength should be $10 \, N/mm^2$ or twice the stress to which the structure will then be submitted, whichever is the greater before striking the formwork. If test cubes are not available, the following table from BS 8110 can be used as a guide where ordinary Portland cement is used.

Location	Surface or air temperature	
	16 °C	7 °C
Vertical formwork	12 hours	18 hours
Slab soffits (props left under)	4 days	6 days
Removal of props	10 days	15 days
Beam soffits (props left under)	10 days	15 days
Removal of props	14 days	21 days

In very cold weather the above minimum periods should be doubled and when using rapid-hardening Portland cement the above minimum periods can generally be halved.

Formwork must be removed slowly, as the sudden removal of the wedges is equivalent to a shock load being placed upon the partly hardened concrete. Materials and/or plant should not be placed on the partly hardened concrete without the engineer's permission. When the formwork has been removed it should be carefully cleaned to remove any concrete adhering to the face before being re-used. If the forms are not required for immediate re-use they should be carefully stored and stacked to avoid twisting.

The method of curing the concrete will depend upon climatic conditions, type of cement used and the average temperature during the curing period. The objective is to allow the concrete to cure and obtain its strength without undue distortion or cracking. It may be necessary to insulate the concrete by covering with polythene sheeting or an absorbent material which is kept damp to control the surface temperature and prevent the evaporation of water from the surface. Under normal conditions using ordinary Portland cement and with an average air temperature of over 10 °C this period would be two days rising to four days during hot weather and days with prolonged drying winds.

PRECAST CONCRETE FRAMES

The overall concept of a precast concrete frame is the same as any other framing material. Single or multi-storey frames can be produced on the skeleton or box frame principle. Single- and two-storey buildings can also be produced as portal frames. Most precast concrete frames are produced as part of a 'system' building and therefore it is only possible to generalize in an overall study of this method of framing.

Advantages

1. Mixing, placing and curing of the concrete carried out under factory-controlled conditions which results in uniform and accurate units. The casting, being an 'off site' activity, will release site space which would have been needed for the storage of cement and aggregates, mixing position, timber store and fabrication area for formwork and the storage, bending and fabrication of the reinforcement.
2. Repetitive standard units reduce costs: it must be appreciated that the moulds used in precast concrete factories are precision made, resulting in high capital costs. These costs must be apportioned over the number of units to be cast.
3. Frames can be assembled on site in cold weather which helps with the planning, programming and progressing of the building operations. This is important to the contractor since delays can result in the monetary penalty clauses, for late completion of the contract, being invoked.
4. In general the frames can be assembled by semi-skilled labour. With the high turnover rate of labour within the building industry operatives can be recruited and quickly trained to carry out these activities.

Disadvantages

1. System building is less flexible in its design concept than purpose-made structures. It must be noted that there is a wide variety of choice of systems available to the designer, so that most design briefs can be fulfilled without too much modification to the original concept.

2. Mechanical lifting plant will be needed to position the units; this can add to the overall contracting costs since generally larger plant is required for precast concrete structures than for *in-situ* concrete structures.
3. Programming may be restricted by controls on delivery and unloading times laid down by the police. Restrictions on deliveries is a point which must be established at the tender period so that the tender programme can be formulated with a degree of accuracy and the cost of any special requirements can be included in the unit rates for pricing.
4. Structural connections between the precast concrete units can present both design and contractual problems. The major points to be considered are protection against weather, fire and corrosion, appearance and the method of construction. The latter should be issued as an instruction to site, setting out in detail the sequence, temporary supports required and full details of the joint.

METHODS OF CONNECTIONS

FOUNDATION CONNECTIONS

Precast columns are connected to their foundations by one of two methods, depending mainly upon the magnitude of the load. For light and medium loads the foot of the column can be placed in a pocket left in the foundation. The column can be plumbed and positioned by fixing a collar around its perimeter and temporarily supporting the column from this collar by using raking adjustable props. Wedges can be used to give added rigidity while the column is being grouted into the pocket (see Fig. 9.4.1). The alternative method is to cast or weld on a base plate to the foot of the column and use holding-down bolts to secure the column to its foundation in the same manner as described in detail for structural steelwork (see Fig. 9.4.1).

COLUMN CONNECTIONS

The main principle involved in making column connections is to ensure continuity and this can be achieved by a variety of methods. In simple connections a direct bearing and grouted dowel joint can be used, the dowel being positioned in the upper or lower column. Where continuity of reinforcement is required the reinforcement from both upper and lower columns is left exposed and either lapped or welded together before completing the connection with *in-situ* concrete. A more complex method is to use a stud and plate connection where one set of threaded bars are connected through a steel plate welded to a set of bars projecting from the lower column; again the connection is completed with *in-situ* concrete. Typical column connections are shown in Fig. 9.4.2. Column connections should be made at floor levels but above the beam connections, a common dimension being 600 mm above structural floor level. The columns can be of single- or multi-storey height, the latter having provisions for beam connections at the intermediate floor levels.

Figure 9.4.1 Precast concrete column to foundation connections

BEAM CONNECTIONS

As with columns, the main emphasis is on continuity within the joint. Three basic methods are used:

1. A projecting concrete haunch is cast on to the column with a locating dowel or stud bolt to fix the beam.

Figure 9.4.2   Precast concrete column connections

2. A projecting metal corbel is fixed to the column and the beam is bolted to the corbel.
3. Column and beam reinforcement, generally in the form of hooks, are left exposed. The two members are hooked together and covered with *in-situ* concrete to complete the joint.

Column connection
bearing plate

Precast concrete
column

Projecting tie bars
for floor units

'U' bolt cast into
column for edge
beam connection

Nut and washer
recess

Tie
bars

PCC edge
beam rebated
to provide
bearing for
floor units

Reinforced
haunch

Dowel

Cross-beam recessed
at end to receive tie
bars and location dowel

Note: Beam recess filled with cement grout to complete connection

PCC column

PCC rebated edge beam

Bearing plate
bolted to bracket

Tie bars for
floor units

Connection
completed with
in-situ concrete

Angle bracket
bolted to column

Alternative 'tee' bars
cast into column and
bolted to projecting
plate cast into beam

Figure 9.4.3 Typical precast concrete beam connections

With most beam-to-column connections lateral restraint is provided by leaving
projecting reinforcement from the beam sides to bond with the floor slab or precast
concrete floor units (see Fig. 9.4.3).

STRUCTURAL STEELWORK FRAMES

Structural steel, as a means of constructing a framed building, has been used since the beginning of the twentieth century and was the major structural material used until the advent of the Second World War, which led to a shortage of the raw material. This shortage led to an increase in the use of *in-situ* and precast concrete frames. Today both systems are used and this means that a comparison must be made before any particular framing medium is chosen. The main factors to be considered in making this choice are:

- Site costs
- Construction costs
- Maintenance costs

Site costs

A building owner will want to obtain a financial return on his capital investment as soon as possible, therefore speed of construction is of paramount importance. The use of a steel or precast concrete frame will enable the maximum amount of prefabrication off site, during which time the general contractor can be constructing the foundations in preparation for the erection of the frame. To obtain the maximum utilization of a site the structure needs to be designed so that the maximum amount of rentable floor area is achieved. Generally, prefabricated section sizes are smaller than comparable *in-situ* concrete members, due mainly to the greater control over manufacture obtainable under factory conditions, and thus these will occupy less floor area.

Construction costs

The main factors are design considerations, availability of labour, availability of materials and site conditions. Concrete is an adaptable material which allows the designer to be more creative than working within the rigid confines of standard steel sections. However, as the complexity of shape and size increases so does the cost of formwork and for the erection of a steel structure skilled labour is required, whereas activities involved with precast concrete structures can be carried out by the more readily available semi-skilled labour working under the direction of a competent person. The availability of materials fluctuates and only a study of current market trends can give an accurate answer to this problem. Site conditions regarding storage space, fabrication areas and manoeuvrability around and over the site can well influence the framing method chosen.

Maintenance costs

These can be considered in the short or long term but it is fair to say that in most framed buildings the costs are generally negligible if the design and workmanship is sound. Steelwork, because of its corrosive properties, will need some form of protective treatment but since most steel structures have to be given a degree of fire resistance the fire protective method may well perform the dual function.

STRUCTURAL STEEL FRAMES

The design, fabrication, supply and erection of a structural steel frame is normally placed in the hands of a specialist subcontractor. The main contractor's task is to provide the foundation bases in the correct positions and to the correct levels with the necessary holding-down fixing bolts. The designer will calculate the loadings, stresses and reactions in the same manner as for reinforced concrete and then select a standard steel member whose section properties meet the design requirements.

Standard hot-rolled sections are given in BS 4 and BS 4848 and in the *Handbook on Structural Steelwork* published jointly by the British Constructional Steelwork Association and the Steel Construction Institute, which provide the following section types:

Universal beams   These are a range of sections supplied with tapered or parallel flanges and are designated by their serial size × mass in kilograms per metre run. To facilitate the rolling operation of universal beam sections the inner profile is a constant dimension for any given serial size. The serial size is therefore only an approximate width and breadth and is given in millimetres.

Joists   These are a range of small size beams which have tapered flanges and are useful for lintels and small frames around openings. In the case of joists the serial size is the overall nominal dimension.

Universal columns   These members are rolled with parallel flanges and are designated in the same manner as universal beams. It is possible to design a column section to act as a beam and conversely a beam section to act as a column.

Channels   These are rolled with tapered flanges and designated by their nominal overall dimension × mass per metre run and can be used for trimming and bracing members or as a substitute for joist sections.

Angles   These are light framing and bracing sections with parallel flanges. The flange or leg lengths can be equal or unequal and the sections are designated by the nominal overall leg lengths × nominal thickness of the flange.

Universal beams   **Universal columns**

Note: Dimensions shown are serial sizes

Joists   **Channels**

Angles

T-bars

Also available as
long stalk T-bars
and as structural
tees cut from UB
and UC sections

Figure 9.5.1   Typical BS 4 and BS 4848 steel sections

T bars   These are used for the same purposes as angles and are available as rolled
sections with a short or long stalk or, alternatively, they can be cut from a standard
universal beam or column section. Designation is given by the nominal overall
breadth and depth × mass per metre run.

Typical standard steel sections are shown in Fig. 9.5.1.

COLD-ROLLED SECTIONS

Cold-rolled steel sections to BS 2994 have become a popular alternative. They make more effective use of steel to permit selection of profiles more appropriate to the design loading. They are particularly useful in lightweight situations where standard hot-rolled sections would be excessive. These include prefabricated construction of modular units for 'slotting' into the main structural frame and for other factory-made units such as site huts.

Development of Zed and Sigma sections have proved very successful alternatives to the traditional use of steel angles for truss purlins. They have many advantages, for example:

- They are more cost effective per unit of weight.
- Deck fixings are shorter and simpler to locate around the upper flange.
- A shear centre is contained within the section to provide greater resistance to twisting.

Several examples of standard cold-rolled sections are shown in Fig. 9.5.2 with Zed and Sigma profiles in Fig. 9.5.3.

CASTELLATED UNIVERSAL SECTIONS

These are formed by flame cutting a standard universal beam or column section along a castellated line; the two halves so produced are welded together to form an open web beam. The resultant section is one and a half times the depth of the section from which it was cut (see Fig. 9.5.4). This increase in depth gives greater resistance to deflection without adding extra weight but will reduce the clear headroom under the beams unless the overall height of the building is increased. Castellated sections are economical when used to support lightly loaded floor or roof slabs and the voids in the web can be used for housing services. With this form of beam the shear stresses at the supports can be greater than the resistance provided by the web; in these cases one or two voids are filled in by welding into the voids metal blanks.

CONNECTIONS

Connections in structural steelwork are classified as either shop connections or site connections and can be made by using bolts, rivets or by welding.

Bolts

Black bolts

These are the cheapest form of bolt available; the black bolt can be either hot or cold forged, the thread being machined onto the shank. The allowable shear

Figure 9.5.2   Cold-rolled steel sections to BS 2994 and non-standard square and rectangular box sections

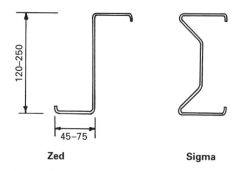

Figure 9.5.3   Cold-rolled steel purlins

Figure 9.5.4 Castellated beams

stresses for this type of bolt are low and therefore they should only be used for end connections of secondary beams or in conjunction with a seating cleat which has been designed and fixed to resist all the shear forces involved. The clearance in the hole for this form of bolt is usually specified as 1.6 mm over the diameter of the bolt. The term black bolts does not necessarily indicate the colour but is the term used to indicate the comparatively wide tolerances to which these products are usually made. Diameters range from 5 to 68 mm inclusive.

Bright bolts

These have a machined shank and are therefore of greater dimensional accuracy, fitting into a hole with a small clearance allowance. The stresses allowed are similar to those permitted for rivets. Bright bolts are sometimes called turned and fitted bolts.

High-strength friction bolts

These are manufactured from high-tensile steels and are used in conjunction with high-tensile steel nuts and tempered washers. These bolts have generally replaced rivets and black bolts for both shop and site connections since fewer bolts are needed and hence the connection size is reduced. The object of this form of bolt is to tighten it to a predetermined shank tension in order that the clamping force thus provided will transfer the loads in the connecting members by friction between the parts and not by shear in, or bearing on, the bolts. Generally a torque-controlled spanner or pneumatic impact wrench is used for tightening; other variations to ensure the correct torque are visual indicators such as a series of pips under the head or washer which are flattened when the correct amount of shank tension has been reached. Nominal standard diameters available are from 12 to 36 mm with lengths ranging from 40 to 500 mm, as recommended in BS 4395.

General

The holes to receive bolts should always be drilled in a position known as the back mark of the section. The back mark is the position on the flange where the removal of material to form a bolt or rivet hole will have the least effect upon the section properties. Actual dimensions and recommended bolt diameters are given in the *Handbook on Structural Steelwork*.

Rivets

Made from mild steel to the recommendations of BS 4620, rivets have been generally superseded by bolted and welded connections for structural steel frames. Rivets are available as either cold or hot forged with a variety of head shapes ranging from an almost semi-circular or snap head to a countersunk head for use when the projection of a snap, universal or flat head would create an obstruction. Small-diameter rivets can be cold driven but the usual practice is to drive rivets while they are hot. Rivets, like bolts, should be positioned on the back mark of the section; typical spacings are $2\frac{1}{2}$ diameters centre to centre and $1\frac{3}{4}$ diameters from the end or edge to the centre line of the first rivet.

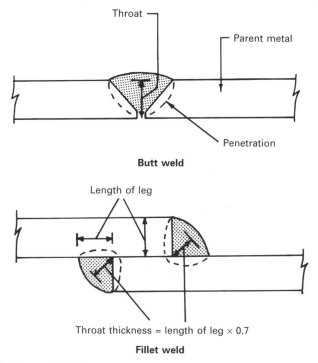

Figure 9.5.5 Types of weld

Welding

Welding is primarily considered as a shop connection since the cost together with the need for inspection, which can be difficult on site, generally makes this method uneconomic for site connections.

The basic methods of welding are oxy-acetylene and electric arc. A blowpipe is used for oxy-acetylene which allows the heat from the burning gas mixture to raise the temperature of the surfaces to be joined. A metal filler rod is held in the flame and the molten metal from the filler rod fuses the surfaces together. For most structural steel applications, the oxy-acetylene method is limited to cutting.

In the alternative method an electric arc is struck between a metal rod connected to a suitable low-voltage electrical supply and the surface to be joined, which must be earthed or resting on an earthed surface. The heat of the arc causes the electrode or metal rod to melt and the molten metal can be deposited in layers to fuse the pieces to be joined together. With electric arc welding the temperature rise is confined to the local area being welded, whereas the oxy-acetylene method causes a rise in metal temperature over a general area.

Welds are classified as either *fillet* or *butt* welds. Fillet welds are used on the edges and ends of members and form a triangular fillet of welding material. Butt welds are used on chamfered end to end connections (see Fig. 9.5.5).

STRUCTURAL STEEL CONNECTIONS

Base connections

These are of one or two forms, the slab or bloom base and the gusset base. In both methods a steel base plate is required to spread the load of the column onto the foundation. The end of the column and the upper surface of the base plate should be machined to give a good interface contact when using a bloom base. The base plate and column can be connected together by using cleats or by fillet welding (see Fig. 9.5.6).

The gusset base is composed of a number of members which reduce the thickness of the base plate and can be used to transmit a high bending moment to the foundations. A machined interface between column and base plate will enable all the components to work in conjunction with one another, but if this method is not adopted the connections must transmit all the load to the base plate (see Fig. 9.5.6). The base is joined to the foundation by holding-down bolts which must be designed to resist the uplift and tendency of the column to overturn. The bolt diameter, bolt length and size of plate washer are therefore important. To allow for fixing tolerances the bolts are initially housed in a void or pocket which is filled with grout at the same time as the base is grouted onto the foundation. To level and plumb the columns steel wedges are inserted between the underside of the base plate and the top of the foundation (see Fig. 9.5.6).

Beam to column connections

These can be designed as simple connections where the whole of the load is transmitted to the column through a seating cleat. This is an expensive method requiring heavy sections to overcome deflection problems. The usual method employed is the semi-rigid connection where the load is transmitted from the beam to the column by means of top cleats and/or web cleats; for ease of assembly an erection cleat on the underside is also included in the connection detail (see Fig. 9.5.7). A fully rigid connection detail, which gives the greatest economy on section sizes, is made by welding the beam to the column (see Fig. 9.5.7). The uppermost beam connection to the column can be made by the methods described above or, alternatively, a bearing connection can be used, which consists of a cap plate fixed to the top of the column to which the beams can be fixed either continuously over the cap plate or with a butt joint (see Fig. 9.5.7).

Column splices

These are made at floor levels but above the beam connections. The method used will depend upon the relative column sections (see Fig. 9.5.8).

Figure 9.5.6 Structural steel column bases

Beam-to-beam connections

The method used will depend upon the relative depths of the beams concerned. Deep beams receiving small secondary beams can have a shelf angle connection whereas other depths will need to be connected by web cleats (see Fig. 9.5.9).

3 mm expansion gap

Universal beam

Universal beams

Site connections

Shop connections

Web cleat

Flange cleat

Simple connection

Site connections

Universal column

Top cleat

Shop connections

Universal beams

Universal beam

Seating or erecting cleat

Web cleats to both sides if required

Semi-rigid connection

Universal column

Fillet weld

Universal beams

Universal beam

Erection cleats

Rigid connection

Figure 9.5.7   Structural steel beam to column connections

Columns with equal sections

Note: For columns of same serial size but of different sections splice is made using 4 No. fish plates fixed on the inside of flanges

Alternative methods for columns of unequal sections

Note: Splices made at floor level but above beam connections

Figure 9.5.8   Structural steel column splices

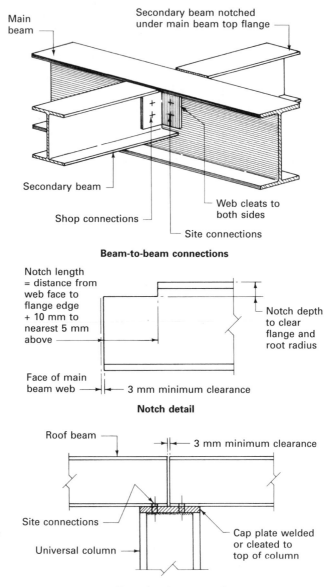

Main beam

Secondary beam notched under main beam top flange

Secondary beam

Shop connections

Web cleats to both sides

Site connections

Beam-to-beam connections

Notch length = distance from web face to flange edge + 10 mm to nearest 5 mm above

Notch depth to clear flange and root radius

Face of main beam web

3 mm minimum clearance

Notch detail

Roof beam

3 mm minimum clearance

Site connections

Universal column

Cap plate welded or cleated to top of column

Beam-bearing connection

Figure 9.5.9   Structural steel beam-to-beam connections

FRAME ERECTION

This operation will not normally be commenced until all the bases have been cast and checked since the structural steelwork contractor will need a clear site for manoeuvring the steel members into position. The usual procedure is to erect two storeys of steelwork before final plumbing and levelling takes place.

The grouting of the base plates and holding-down bolts is usually left until the whole structure has been finally levelled and plumbed. The grout is a neat cement or cement/sand mixture depending on the gap to be filled:

- 12 to 25 mm gap – stiff mix of neat cement.
- 25 to 50 mm gap – fluid mix of 1:2 cement/sand and tamped.
- Over 50 mm gap – stiff mix of 1:2 cement/sand and rammed.

With large base plates a grouting hole is sometimes included but with smaller plates three sides of the base plate are sealed with puddle clay, bricks or formwork and the grout introduced through the open edge on the fourth side. To protect the base from corrosion it should be encased with concrete up to the underside of the floor level giving a minimum concrete cover of 75 mm to all the steel components.

FIRE PROTECTION OF STEELWORK

The Building Regulations, Approved Document B, gives minimum fire resistance periods for steel structures and components according to a building's purpose group and its top floor height above ground. The traditional method of fire protection is to encase the steel section with concrete. This requires formwork and will greatly add to the structural dead load. 50 mm cover will provide up to 4 hours fire protection, whereas the more attractive hollow protection from brickwork will require at least 100 mm thickness for equivalent protection and will occupy more space. A number of lighter weight and less bulky sheet materials based on plasterboard can be used to 'box' steel columns and beams, but due to their porous nature are limited to interior use. They are also less practical where more than 2 hours fire protection is required. Some basic examples of solid and hollow protection to steelwork are shown in Fig. 9.5.10. For additional details and specific treatment of the subject of fire protection consult the complementary volume, *Advanced Construction Technology*.

50 mm = 2 h protection
100 mm = 4 h protection

$25 \text{ mm} = \frac{1}{2} \text{ h protection}$
$50 \text{ mm} = 4 \text{ h protection}$

Figure 9.5.10   Fire protection to steel columns

CLADDINGS

Claddings to buildings can be considered under two classifications:

- Claddings fixed to a structural backing.
- Claddings to framed structures.

CLADDINGS FIXED TO A STRUCTURAL BACKING

Materials used in this form of cladding are generally considered to be small unit claddings and are applied for one of two reasons. If the structural wall is unable to provide an adequate barrier to the elements, a covering of small unit claddings will generally raise the wall's resistance to an acceptable level. Alternatively, small unit claddings can be used solely as a decorative feature, possibly to break up the monotony of a large plain area composed of a single material.

The materials used are tiles, slates, shingles, timber boarding, plastic boards and stone facings. The general method of fixing these small units is to secure them to timber battens fixed to the structure backing. Stone and similar facings, however, are usually secured by special mechanical fixings as described later when considering claddings to framed structures.

TILE HANGING

The tiles used in the hanging can be ordinary plain roofing tiles or, alternatively, a tile of the same dimensions but having a patterned bottom edge solely for a decorative appearance. The tiles are hung and fixed to tiling battens although nibless tiles fixed directly to the backing wall are sometimes used (see Fig. 9.6.1). The battens should be impregnated to prevent fungi and insect attack so that their anticipated life is comparable to that of the tiles. Each tile should be twice nailed to its support batten with corrosion-resistant nails of adequate length.

The general principles of tile hanging are similar to those of double-lap roof tiling and the gauge is calculated in the same manner. The minimum lap recommended is 40 mm which would give a gauge of 112.5 mm using a standard 265 long tile.

Eaves tile under bottom course

Pointed pattern **Fishtail pattern** **Club pattern**

For details of roof construction and tiling see Chapter 6.2

Rainwater gutter

Fascia board

Ventilate soffit board if roof is insulated

Cover moulding fixed to soffit

Lead or similar flashing fixed behind top batten

Top course of nibless tiles

Brick or block backing wall in solid or cavity construction complying with the requirements of the Building Regulations for structural stability and thermal insulation

265 × 165 plain or patterned tiles each twice nailed with galvanized nails to 38 × 25 softwood battens and felt breather membrane fixed to backing to a 110 mm gauge

76 × 65 softwood tilting fillet

An alternative associated with older buildings: brick backing in rat trap bond clad with nibless tiles fixed directly into mortar joints

Figure 9.6.1 Vertical tile hanging – typical details 1

Alternative external angle treatments
(internal angles treatments similar)

- Backing wall
- 38 × 25 battens to 110 mm gauge
- Code 4 lead soakers
- Plain tiles
- Mitred plain tiles and mitred tiles and a half
- Special one piece corner tiles
- Plain tiles

Typical abutment detail

- Backing wall
- Code 5 lead stepped flashing turned into horizontal joints wedged and pointed in cm. mt. (1:3)
- Plain tiles hung on 38 × 25 battens

Typical opening details

- Window sill
- Code 5 lead flashing
- Vertical tiling
- Window frame jamb
- cm. mt. (1:3) pointing

Figure 9.6.2 Vertical tile hanging – typical details 2

A gauge dimension of 112.5 mm is impracticable and therefore a gauge of 110 mm would be usual. Typical details of top edge finishes, bottom edge finishes, corners and finishes at windows are shown in Figs 9.6.1 and 9.6.2. It should be noted that if the structural backing is of timber framing, an impervious tiling felt breather membrane should be placed over the framing immediately underneath the battens to prevent any moisture which is blown in between the tiles from having adverse effects upon the structure. As a matter of course, it is good practice to underlay the tile battens with a breather membrane, whatever the wall structure material. In this situation building paper is not considered to be a suitable substitute. The application of slates as a small unit hung cladding follows the principles outlined above for tile hanging.

TIMBER CLADDINGS

Timber claddings are usually in the form of moulded or shaped boards fixed to battens as either a horizontal or vertical cladding; typical examples are shown in Fig. 9.6.3. Timber claddings will require regular maintenance to preserve their resistance to the elements. Softwoods are generally painted and will need repainting at intervals of three to five years according to the exposure. Hardwoods are sometimes treated with a preservative and left to bleach naturally; the preservative treatment needs to be carried out at two- to five-year intervals. Western red cedar is a very popular wood for timber cladding since it has a natural immunity to insect and fungi attack under normal conditions. It also has a pleasing natural red/brown colour which can be maintained if the timber is coated with a clear sealer such as polyurethane; however, it will bleach to a grey/white colour if exposed to the atmosphere. Plastic boards are a substitute for timber and are fixed in a similar manner.

CLADDINGS TO FRAME STRUCTURES

The methods available to clad a frame structure are extensive and include panels of masonry constructed between the columns and beams, light infill panels of metal or timber, precast concrete panels and curtain walling which completely encloses the structure. This volume includes panel walls and the facings which can be attached to them. The other forms of cladding are considered in the complementary volume, *Advanced Construction Technology*.

BRICK PANEL WALLS

These are non-loadbearing walls which must fulfil the following requirements:

1. Adequate resistance to the elements.
2. Have sufficient strength to support their own weight plus any attached finishes.
3. Strong enough to resist both positive and negative wind pressures.
4. Provide the required thermal and sound insulation.
5. Provide the required fire resistance.
6. Have adequate durability.

Brick panel walls are constructed in the same manner as ordinary solid or cavity walls and any openings for windows or doors are formed by traditional

Brick or block backing wall

Breather membrane

38 × 25 battens at 450 to 600 c/c

Feather edge boarding with 32 mm overlap

150

20

10

Tilting fillet to bottom edge

Brick or block backing wall

Breather membrane

38 × 25 battens at 450 to 600 c/c

Cedar or deal shiplap boarding

150

20

Closing fillet to bottom edge

Typical horizontal claddings

Brick or block backing wall

32 × 20 cover moulding

150 × 20 square edge boards

Breather membrane

38 × 25 battens at 450 to 600 c/c

Cedar or deal matchboarding

Typical vertical cladding

Figure 9.6.3   Timber wall claddings

methods. The panels must be supported at each structural floor level and tied to the structure at the vertical edges. Projection of the panel in front of the structural members is permissible, providing such overhangs do not impair the stability of the panel wall; acceptable limits are shown in Fig. 9.6.4. The top edge of the panels should not be pinned rigidly to the frame since the effect of brick panel expansion together with frame shrinkage may cause cracking and failure of the brickwork. A compression joint should therefore be formed between the top edge of the panel and the underside of the framing member at each floor level (see Fig. 9.6.4).

Two methods of tying the panel to the vertical structural members are in common use:

1. Butterfly wall tiles are cast into the column and built into the brick joints at four-course intervals.
2. Galvanized pressed steel dovetail slots are cast into the column and dovetail anchors are used to form the tie (see Fig. 9.6.4).

The second method gives greater flexibility with the location and insertion of adequate ties but is higher in cost.

Location of masonry and its interaction with the structural frame may also be achieved with the support framework shown in Fig. 9.6.5. A galvanized steel dove-tailed profile is cast into the beam face to create a recess for fixings. It supports a continuous steel angle, which in turn supports the outer leaf of masonry from the edge beam. A sliding anchor system is also bolted to the recess to retain both elements of the cavity wall.

Facings to brick panel walls

Any panel wall must have an acceptable and durable finish; this can be achieved by using facing bricks with a neat pointed joint or by attaching to the face of a panel of common bricks a stone or similar cladding. Suitable materials are natural stone, artificial stone, reconstructed stone and precast concrete of small units up to 1 m^2 and with a thickness related to the density of the material. Dense materials such as slate and marble need only be 40 mm thick, whereas the softer stones such as sandstone and limestone should be at least 75 mm thick.

Two major considerations must be taken into account when deciding on the method to be used to fix the facings to the brick backing:

1. Transferring the load to the structure.
2. Tying back the facing units.

The load of the facings can be transferred by using bonder stones or support corbels at each floor level, which should have a compression joint incorporated in the detail for the same reasons given above when considering brick panels (see Fig. 9.6.6).

Maximum overhang for panel walls

Typical beam facing details

Figure 9.6.4 Brick panel walls

Figure 9.6.5   Masonry cladding support system

The tying back of the facings is carried out by various metal fixing devices called *cramps* which should be of a non-ferrous metal such as gunmetal, copper, phosphor bronze or stainless steel. To avoid the problem of corrosion caused by galvanic action between dissimilar metals a mixture of fixing materials should not be used. Typical examples of fixings and cramps for thick and thin facings are shown in Fig. 9.6.6.

To provide for plumbing and alignment a bedding space of 12–15 mm should be left between the face of the brick panel and the back of the facing. Dense facings such as marble are usually bedded on a series of cement mortar dabs, whereas the more porous facings are usually placed against a solid bed which ensures that any saturation which occurs will be uniform over the entire face.

[*Note*: With all forms of masonry cladding to structural frames, due regard must be given to provision of adequate thermal insulation and avoidance of thermal bridging. The details shown in this chapter indicate the principles of various cladding techniques, but care must be taken to ensure that the structure and claddings are selected appropriate to the building's location and function, with regard for local building regulations.]

Cramps for horizontal joints

Half-twist cramps for vertical joints

Typical cladding fixing details

Corbel support

Thin cladding fixing

Figure 9.6.6 Cladding fixings

STEEL ROOF TRUSSES AND COVERINGS

The function of any roof is to provide a protective covering to the upper surface of the structure. By virtue of its position a roof is subjected to the elements to a greater degree than the walls, therefore the durability of the covering is of paramount importance. The roof structure must have sufficient strength to support its own weight, the load of the coverings together with any imposed loadings such as snow and wind pressures without collapse or excessive deflection.

Roofs can be considered as:

- Short span.
- Medium span.
- Long span.

Short span

Up to 7.000 m, generally of traditional timber construction with a flat or pitched profile. Flat roofs are usually covered with a flexible sheet material whereas pitched roofs are generally covered with small units such as tiles or slates.

Medium span

7.000 to 24.000 m, except where reinforced concrete is used; the usual roof structure for a medium span is a truss or lattice of standard steel sections supporting a deformed sheeting such as corrugated asbestos cement or a structural decking system.

Long span

Over 24.000 m; these roofs are generally designed by a specialist using girder, space deck or vaulting techniques and are beyond the scope of a basic technology course.

STEEL ROOF TRUSSES

This form of roof structure is used mainly for short- and medium-span single-storey buildings intended for industrial or recreational use. A steel roof truss is a plane frame consisting of a series of rigid triangles composed of compression and tension members. The compression members are called *rafters* and *struts*, whereas the tension members are termed *ties*. Standard mild steel angles complying with the recommendations of BS 4848 are usually employed as the structural members and these are connected together, where the centre lines converge, with flat-shaped plates called *gussets*. They can be riveted, bolted or welded together to form a rigid triangulated truss; typical arrangements are shown in Fig. 9.7.1. The internal arrangement of the struts and ties will be governed by the span. The principal or rafter is divided into a number of equal divisions which locates the intersection point for the centre line of the internal strut or tie.

Angle purlins are used longitudinally to connect the trusses together and provide the fixing medium for the roof covering. It is the type of covering chosen which will determine the purlin spacing and the pitch of the truss; ideally the purlins should be positioned over the strut or tie intersection points to avoid setting up local bending stresses in the rafters. Purlins are connected to cleats attached to the backs of the rafters; alternatively a zed section can be used, thus dispensing with the need for a fixing cleat. Steel roof trusses are positioned at 3.000 to 7.500 m centres and supported by capped universal columns or bolted to padstones bedded on to brick walls or piers. The main disadvantage of this form of roofing is the large and virtually unusable roof space. Other disadvantages are the frequent necessity of painting the members to inhibit rust and that the flanges of the angles provide an ideal ledge on which dust can accumulate. Typical steel roof truss details are shown in Fig. 9.7.2.

Suitable truss and girder arrangements can be fabricated from welded steel tubes which are lighter in weight, cleaner in appearance, have less surface area on which to collect dust and therefore less surface area to protect with paint.

COVERINGS

The basic requirements for covering materials to steel roof trusses are:

1. Sufficient strength to support imposed wind and snow loadings.
2. Resistance to the penetration of rain, wind and snow.
3. Low self-weight, so that supporting members of an economic size can be used.
4. Acceptable standard of thermal insulation if habitable or occupational accommodation requiring space heating.
5. Acceptable fire resistance and resistance to spread of flame.
6. Durable to reduce the maintenance required during the anticipated life of the roof.

Most of the materials used for covering a steel roof structure have poor thermal insulation properties and therefore a combination of materials is required if heat

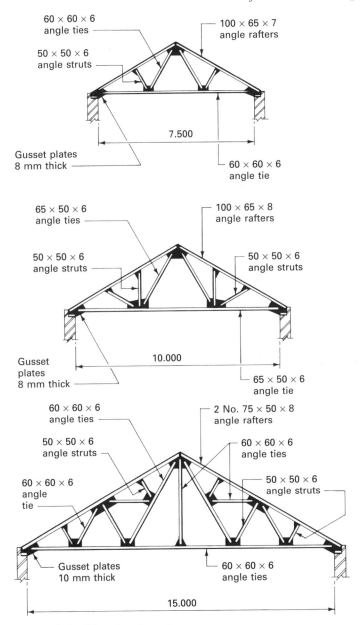

60 × 60 × 6
angle ties

100 × 65 × 7
angle rafters

50 × 50 × 6
angle struts

7.500

Gusset plates
8 mm thick

60 × 60 × 6
angle tie

65 × 50 × 6
angle ties

100 × 65 × 8
angle rafters

50 × 50 × 6
angle struts

50 × 50 × 6
angle struts

Gusset
plates
8 mm thick

10.000

65 × 50 × 6
angle tie

60 × 60 × 6
angle ties

2 No. 75 × 50 × 8
angle rafters

50 × 50 × 6
angle struts

60 × 60 × 6
angle ties

60 × 60 × 6
angle
tie

50 × 50 × 6
angle struts

Gusset plates
10 mm thick

60 × 60 × 6
angle ties

15.000

Figure 9.7.1 Typical mild steel angle roof trusses

loss, or gain, is to be controlled to satisfy legal requirements or merely to conserve the fuel required to heat the building.

Corrugated roofing materials, if correctly applied, will provide a covering which will exclude the rain and snow but will allow a small amount of wind penetration

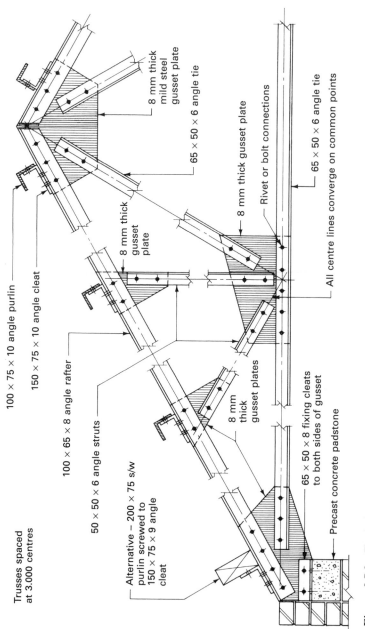

Trusses spaced
at 3.000 centres

100 × 75 × 10 angle purlin

150 × 75 × 10 angle cleat

100 × 65 × 8 angle rafter

50 × 50 × 6 angle struts

Alternative – 200 × 75 s/w
purlin screwed to
150 × 75 × 9 angle
cleat

8 mm thick
mild steel
gusset plate

65 × 50 × 6 angle tie

8 mm thick
gusset
plate

8 mm thick gusset plate

Rivet or bolt connections

All centre lines converge on common points

65 × 50 × 6 angle tie

8 mm
thick
gusset plates

65 × 50 × 8 fixing cleats
to both sides of gusset

Precast concrete padstone

Figure 9.7.2   Typical medium span mild steel roof truss unless the end laps are sealed with two ribbons of silicon or butyl mastic sealant. These coverings are designed to support normal snow loads but are not usually strong enough to support the weight of operatives and therefore a crawling ladder or board should be used.

Owing to the poor thermal insulation value of these roofing materials there is a risk of condensation occurring on the internal surface of the sheets. This risk can be reduced by using a suitable lining at rafter level or by a ceiling at the lower tie level. Unless a vapour control layer is included on the warm side of the lining water vapour may pass through the lining and condense on the underside of the covering material, which could give rise to corrosion of the steel members. An alternative method is to form a 25 mm wide ventilation cavity between the lining and the covering.

Galvanized corrugated steel sheets

These sheets are often referred to as 'corrugated iron' and have been widely used for many years for small industrial and agricultural buildings; they can also be used as a cladding to post and rail fencing. They are generally made to the recommendations of BS 3083 which specifies the sizes, number of corrugations and the quality of the zinc coating or galvanizing.

The pitch of the corrugations, which is the distance between centres of adjacent corrugations, is 75 mm with 7, 8, 9, or 10 corrugations per sheet width with lengths ranging from 1.500 to 3.600 m. A wide range of fittings for ridge, eaves and verge treatments are available. The sheets are secured to purlins with hook bolts, drive screws or nuts and bolts in a similar manner to that detailed for asbestos cement sheets in Fig. 9.7.5. The purlins are spaced at centres from 1.500 to 3.000 m according to the thickness of the sheeting being used. To form a weather-tight covering the sheets are lapped at their ends and sides according to the pitch and exposure conditions:

- **End laps** – up to 20° pitch 150 mm minimum and sealed with a butyl or silicon sealant.
- **Side laps** – formed on edge away from the prevailing wind with a 1½ corrugation lap for conditions of normal exposure and two corrugation lap for conditions of severe exposure.

When fixed, galvanized corrugated steel sheets form a roof which is cheap to construct, strong, rigid and non-porous. On exposure the galvanized coating oxidizes, forming a thin protective film which is easily broken down by acids in the atmosphere. To extend the life of the sheeting it should be regularly coated with paint containing a pigment of zinc dust, zinc oxide, calcium plumbate or zinc chromate. The use of these paints will eliminate the need for an application of mordant solution to provide a key. When laying new sheeting it is advisable to paint under the laps before fixing since the overlap is very vulnerable to early corrosion.

The main disadvantages of this form of roof covering are:

1. Poor thermal insulation properties – 8.6 W/m² K which can be reduced by using a 12 mm insulation fibre board in conjunction with a 25 mm cavity to 1.7 W/m² K.
2. Although a non-combustible material, galvanized corrugated steel sheets tend to buckle under typical fire conditions.
3. Inclined to be noisy during rain which produces a 'drumming' sound.

Corrugated asbestos cement sheets

This was the major covering material used for cladding steel roof structures and is made from asbestos fibres and cement in the approximate proportions of 1:7 together with a controlled amount of water. Corrugated asbestos cement sheets were produced to the recommendations of BS 690: Part 3 (now withdrawn), together with a wide range of fittings for the ridge, eaves and verge which are used in conjunction with the various profiles produced (see Figs 9.7.3 and 9.7.4).

Concern with the health risk attached to the manufacture and use of asbestos-based products has led to the development and production of alternative fibre-based materials including profiles to match the corrugated asbestos cement sheets. These now conform to BS EN 494.

Fibre cement profiled sheets

This alternative material to asbestos cement has been developed to meet the same technical specification with a similar low maintenance performance. Fibre cement sheets are made by combining natural and synthetic non-toxic fibres and fillers with Portland cement and, unlike asbestos cement sheets which are rolled to form the required profile, these sheets are pressed over templates. The finished product has a natural grey colour but sheets with factory applied surface coatings are available.

The sheets and fittings are fixed through the crown of the corrugation using either shaped bolts to steel purlins or drive screws to timber purlins. At least six fixings should be used for each sheet and to ensure that a weather-tight seal is achieved at the fixing positions a suitable felt or lead pad with a diamond-shaped curved washer can be used. Alternatively a plastic sealing washer can be employed (see Figs 9.7.5 and 9.7.6). The sheets can be easily drilled for fixings which should be 2 mm larger in diameter than the fixing and sited at least 40 mm from the edge of the sheet. Side laps should be positioned away from the prevailing wind and end laps on low pitches should be sealed with a mastic or suitable preformed compressible strip.

The *U* value of fibre cement sheets is high, generally about 6.0 W/m² K, therefore if a higher degree of thermal resistance is required it will be necessary to use a system of underlining sheets with an insulating material sandwiched between the profile and underlining sheet.

Minimum end lap
for all types = 150

Class 1: min. pitch 10°; max. purlin spacing 900 (symmetrical)

Class 2: min. pitch 10°; max. purlin spacing 1.400 (symmetrical)

Class 2: min. pitch 10°; max. purlin spacing 1.400 (asymmetrical)

Class 3: min. pitch 4°; max. purlin spacing 1.680 (asymmetrical)

Class 4: min. pitch 4°; max. purlin spacing 1.980 (symmetrical)

Lengths	
Class 1	1.225 to 3.050
Class 2	1.525 to 3.050
Class 3	1.675 to 3.050
Class 4	1.825 to 3.050

Figure 9.7.3 Typical corrugated sheet profiles

Figure 9.7.4   Typical fittings for Class 2 sheeting

Aluminium sheeting

This form of roof covering is available in a corrugated or troughed profile usually conforming to the requirements of BS 4868. The sheets are normally made from an aluminium–manganese alloy resulting in a non-corrosive, non-combustible lightweight sheet (2.4 to 5.0 kg/m^2). Aluminium sheeting oxidizes on the surface to form a protective film upon exposure to the atmosphere and therefore protective treatments are not normally necessary. Fixings of copper or brass should not be

150 min.

50 min.

Sheeting

8 mm dia. galvanized
hook bolt with
diamond washer
and square nut

Angle purlin

Plastic protective cap
to bolt and nut

Diamond washer
and pad

Sheeting

8 mm dia. galvanized
'zed' type hook
bolt

'Zed' beam purlin

Diamond
washer and
square nut

150 min.

50 min.

Sheeting

8 mm dia. galvanized
'J' bolt

Tubular purlin

Galvanized spring
head screw with
washered head and
square twisted shank

Galvanized drive
screw with cone
head and twisted
shank

Fixings for timber purlins

Figure 9.7.5   Typical roof sheeting fixings

used since the electrolytic action between the dissimilar materials could cause harmful corrosion, and where the sheets are in contact with steelwork the steel members should be painted with at least two coats of zinc chromate or bituminous paint.

The general application in design and construction of an aluminium covering is similar to that described and detailed for fibre cement sheeting. A wide range of

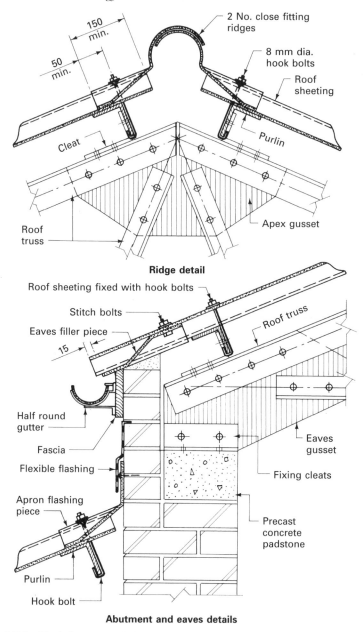

Figure 9.7.6   Typical corrugated sheet roofing details

fittings are available and like the asbestos cement sheets can be fixed with hook bolts, bolts and clips or special shot fasteners. The sheets are intended for a 15° pitched roof with purlins at 1.200 m centres for the 75 mm corrugated profile and at 2.700 m centres for the trough profiles. Laps should be 1½ corrugations for the side lap or 45–57 mm for trough sheets with a 150 mm minimum end lap for all profiles.

Figure 9.7.7 Profiled sheet metal roof covering

Aluminium (BS EN 485) and coated steel (BS EN 10147)

Profiled sheeting

These are produced in a trapezoidal profile in varying depth and pitch dimensions depending on the source of manufacture. Surface treatment may be galvanized for steel, painted or more commonly plastic coated. Various types of plastic coating are possible, including acrylic, polyester and resins. All can be colour pigmented to provide a wide range of choice to suit all types of buildings, including schools, hospitals, workshops, warehousing, offices, etc. Fixings can be through the valley for direct connection to the purlin with self-drilling and tapping screws shown in Fig. 9.7.7. Crown fixing, also shown, has self-tapping screws in pre-drilled holes with a butyl mastic (or equivalent) sealant to prevent ingress of rainwater.

The single skin corrugated/profiled systems of roof covering shown in this chapter are limited to unheated structures such as warehousing, workshops, garages and loading bays. Current thermal insulation legislation for industrial and commercial buildings in the UK requires a substantial thickness of insulation material integral with the roof construction. This may be sandwiched between two layers of profiled covering or incorporated above a ceiling lining. More information on this and the topic of thermal insulation to non-dwellings is to be found in the associated volume, *Advanced Construction Technology*.

Tiles or slate coverings

The traditional tile or slate roof is not limited to dwellings. Many industrial and commercial premises with medium/large span steel roof trusses are finished in these materials. Figure 9.7.8 shows a method of fixing to standard T- or L-shaped steel rafters.

Figure 9.7.8   Slate covering

PART 10

SERVICES

DOMESTIC WATER SUPPLY

An adequate supply of water is a basic requirement for most buildings for reasons of personal hygiene or for activities such as cooking and manufacturing processes. In most areas a piped supply of water is available from the local utility company's water main.

Water is, in the first instance, produced by condensation in the form of clouds and falls to the ground as rain, snow or hail; it then becomes either surface water in the form of a river, stream or lake, or percolates through the subsoil until it reaches an impervious stratum, or is held in a water-bearing subsoil.

The water authority extracts water from surface storage reservoirs, rivers or through boreholes into water-bearing subsoil. Thereafter it is filtered, chlorinated, aerated, and treated with fluoride for human consumption before distribution through a system of mains.

The water company's mains are laid underground at a depth where they will be unaffected by frost or traffic movement. The layout of the system is generally a circuit with trunk mains feeding a grid of subsidiary mains for distribution to specific areas or districts. The materials used for main pipes are cast iron and uPVC which can be tapped while under pressure; a plug cock is inserted into the crown of the mains pipe to provide the means of connecting the communication pipe to supply an individual building.

Terminology

- **Main**   A pipe for the general conveyance of water as distinct from the conveyance to individual premises.
- **Service**   A system of pipes and fittings for the supply and distribution of water in any individual premises.
- **Service pipe**   A pipe in a service which is directly subject to pressure from a main, sometimes called the rising main, inside the building.
- **Communication pipe**   That part of the service pipe which is vested in the water undertaking.
- **Distribution pipe**   Any pipe in a service conveying water from a storage cistern.

461

- **Cistern**   A container for water in which the stored water is under atmospheric pressure.
- **Storage cistern**   Any cistern other than a flushing cistern.
- **Tank**   A rectangular vessel completely closed and used to store water.
- **Cylinder**   A closed cylindrical tank.

COLD WATER SUPPLY

The water company will provide from their mains tapping plug cock a communication pipe to a stop valve and protection chamber just outside the boundary; this is a chargeable item to the building owner. The goose neck bend is included to relieve any stress likely to be exerted on the mains connection.

A service pipe is taken from this stop valve to an internal stop valve, preferably located just above floor level and housed under the sink unit. The stop valve should have a drain-off valve incorporated in it, or just above it, so that the service pipe or rising main can be drained.

Care must be taken when laying a service pipe that it is not placed in a position where it can be adversely affected by frost, heavy traffic or building loads. A minimum depth of 750 mm is generally recommended for supplies to domestic properties; where the pipe passes under a building it should be housed in a protective duct or pipe suitably insulated within 750 mm of the floor level (see Fig. 10.1.1).

Suitable materials for service pipes are copper, PVC, polythene and galvanized steel. Copper service pipes can be laid on and covered by a layer of sand to prevent direct contact with the earth or alternatively wrapped with a suitable proprietary insulating material. Plastic-coated copper pipes are also available for underground pipework. Steel pipes should have a similar protection but plastic pipes are resistant to both frost and corrosion and have largely superseded metal pipes in this situation.

WATER METER

These are a standard installation for all new buildings and conversions within buildings, as a resource conservation measure. Existing premises may have meters installed at the water authority's discretion. Where practicable, the meter is located on the service pipe, in a small compartment below ground and just inside the property boundary. If this proves impossible due to lack of space, the meter may be positioned at the base of the rising main.

DIRECT COLD WATER SUPPLY

In this system the whole of the cold water to the sanitary fittings is supplied directly from the service pipe. The direct system is used mainly in northern districts where large, high-level reservoirs provide a good mains supply and

Cast iron surface box BS 5834

Site boundary

Drain-off valve

Stop valve

Goose neck
Plug cock

Stop valve guard BS 5834

750 min.

Service pipe insulated within drain pipes

Pipe end sealed

Stop valve

Company's main

Sealed with plastic material

75 mm dia. protecting drain pipes

12 mm dia. service pipe at right angles to company's main

Typical cold water service layout

Coupling nut

Friction locking ring

Copper pipe

Copper pipe

Copper coupling fitting

End of pipe swaged

Typical manipulative compression joint

Coupling nut

Copper pipe

Copper pipe

Copper coupling fitting

Tapered compression ring

Typical non-manipulative compression fitting

Copper fitting

Solder retaining groove

Copper pipe

Copper pipe

Fine solder

Typical capillary joint

Figure 10.1.1 Supply pipe and copper pipe joint details

pressure. With this system only a small cold water storage cistern to feed the hot water tank is required; this can usually be positioned below the roof ceiling level giving a saving on pipe runs to the roof space and eliminating the need to insulate the pipes against frost (see Fig. 10.1.2).

Direct system of cold water supply

Indirect system of cold water supply

Note: All pipe sizes shown are standard copper outside diameters

Figure 10.1.2   Cold water supply systems

Another advantage of the direct system is that drinking water is available from several outlet points. The main disadvantage is the lack of reserve should the mains supply be cut off for repairs; also there can be a lowering of the supply during peak demand periods.

When sanitary fittings are connected directly to a mains supply there is always a risk of contamination of the mains water by back siphonage. This can occur if there is a negative pressure on the mains and any of the outlets are submerged below the water level, such as a hand spray connected to the taps.

INDIRECT COLD WATER SUPPLY

In this system all the sanitary fittings, except for a drinking water outlet at the sink, are supplied indirectly from a cold water storage cistern positioned at a high level, usually in the roof space. This system requires more pipework but it gives a reserve supply in case of mains failure and it also reduces the risk of contamination by back siphonage (see Fig. 10.1.2). It should be noted that the local water authority determines the system to be used in the area.

PIPEWORK

Any of the materials which are suitable for the service pipe are equally suitable for distribution pipes and the choice is very often based on individual preference, initial costs and possible maintenance costs.

Copper pipes

Copper pipes have a smooth bore giving low flow resistance, they are strong and easily jointed and bent. Joints in copper pipes can be made by one of three methods:

- **Manipulative compression joint** The end of the pipe is manipulated to fit into the coupling fitting by means of a special tool. No jointing material is required and the joint offers great resistance to being withdrawn. It is usually a by-law requirement that this type of joint is used on service pipes below ground.
- **Non-manipulative compression joint** No manipulation is required to the cut end of the pipe; the holding power of the joint relies on the grip of a copper cone, ring or olive within the joint fitting.
- **Capillary joint** The application of heat makes the soft solder contained in a groove in the fitting flow around the end of the pipe which has been cleaned and coated with a suitable flux to form a neat and rigid joint.

Typical examples of copper pipe joints are shown in Fig. 10.1.1.

Steel pipes

Steel pipes for domestic water supply can be obtained as black tube, galvanized or coated and wrapped for underground services. The joint is usually made with a tapered thread and socket fitting, and to ensure a sound joint, stranded hemp and jointing paste should be wrapped around the thread; alternatively a non-contaminating polytetrafluoroethylene (PTFE) seal tape can be used (see Fig. 10.1.3).

Polythene pipe

Polythene pipe is very light in weight, easy to joint, non-toxic and is available in long lengths, which gives a saving on the number of joints required. Jointing of polythene pipes is generally of the compression type using a metal or plastic liner to the end of the tube (see Fig. 10.1.3). To prevent undue sagging polythene pipes should be adequately fixed to the wall with saddle clips, recommended spacings are fourteen times the outside diameter for horizontal runs and twenty-four times outside diameter for vertical runs.

Unplasticized PVC (uPVC)

This is plastic pipe for cold water services which is supplied in straight lengths up to 9,000 mm long and in standard colours of grey, blue and black. Blue is the water authorities' preferred colour, for easy identification when buried. Jointing can be by a screw thread but the most common method is by solvent welding. This involves cleaning and chamfering the end of the pipe which is coated with the correct type of adhesive and pushed into a straight coupling which has also been similarly coated. The solvent will set within a few minutes but the joint does not achieve its working strength for twenty-four hours. Heat fusion is also used, but mainly for the larger diameter water authority mains. Here the pipe spigot and the coupling are heated with circular dies to melt the uPVC and effect a push fit weld.

COLD WATER STORAGE CISTERNS

The size of cold water storage cisterns for dwelling houses will depend upon the reserve required and whether the cistern is intended to feed a hot water system. Minimum actual capacities recommended in model water by-laws are 115 litres for cold water storage only and 230 litres for cold and hot water services.

Cisterns should be adequately supported and installed in such a position to give reasonable access for maintenance purposes. The cistern must be installed so that its outlets are above the highest discharge point on the sanitary fittings since the flow is by gravity. If the cistern is housed in the roof the pipes and cistern should be insulated against possible freezing of the water; preformed casings of suitable materials are available to suit most standard cistern sizes and shapes. The inlet and outlet connections to the cistern should be on opposite sides to prevent stagnation of water. A securely fitting cover must be provided to prevent ingress of dust, dirt and insects. To prevent a vacuum occurring as water is drawn, the cover is fitted with a screened vent. It also has a moulded sleeve/boss for adapting the hot water expansion pipe. The overflow or warning pipe is also fitted with a filter. A typical cistern installation is shown in Fig. 10.1.3.

Filtered vent (0.65 mm max. opening)

Insulated rising main

Securely fitted cover

Expansion pipe sleeve

Service valve

20 min.
25 min.
50

Filtered 22 mm warning pipe

Float valve

Insulation jacket

Boarded platform

100 × 50 bearers

Ceiling joist

Insulation

22 mm cold feed

15 mm rising main

Typical cold water storage cistern

Piston

Cap

Float arm

Lock nut

Outlet

Back nut

Piston float valve

Galvanized steel socket

Steel pipe

Tapered thread

Steel pipe joint

Nylon nozzle

Rubber diaphragm

Body

Plunger

Lock nut

Float arm

End plate

Back nut

Cap

Outlet preferred at top of body, with a small pipe to direct flow downwards

Diaphragm float valve

Polythene pipe

Copper fitting

Copper liner

Cone wedge

Polythene pipe joint

Figure 10.1.3 Cisterns, float valves and joints

Plastic cisterns have many advantages over the traditional galvanized mild steel cisterns; they are non-corrosive, rot proof, frost resistant and have good resistance to mechanical damage. Materials used are polythene, polypropylene and glass fibre: these cisterns are made with a wall thickness to withstand the water pressure and have an indefinite life. Some forms of polythene cisterns can be distorted to enable them to be passed through an opening of 600×600 mm which is a great advantage when planning access to a roof space. However, it is always better to deposit the cistern within the roof structure during construction, rather than have access difficulties later!

Float valves

Every pipe supplying a cold water storage cistern must be fitted with a float valve to prevent an overflow. The float valve must be fitted at a higher level than the overflow to prevent it becoming submerged and creating the conditions where back siphonage is possible. A float valve is designed to automatically regulate the supply of water by a floating ball closing the valve when the water reaches a predetermined level.

Two valves are in common use for domestic work, namely the piston valve and the diaphragm valve. The piston valve has a horizontal piston or plunger which closes over the orifice of a diameter to suit the pressure; high-, medium- and low-pressure valves are available (see Fig. 10.1.3). The diaphragm valve closes over an interchangeable nylon nozzle orifice. This type of valve is quieter in operation, easily adjustable and less susceptible to the corrosion trouble caused by a sticking piston – this is one of the problems that can be encountered with the piston valve (see Fig. 10.1.3). Diaphragm valves now have the outlet at the top to increase the air gap between outlet and water level, thus reducing the possibility of back siphonage if the cistern water level were to rise excessively.

HOT WATER SUPPLY

The supply of hot water to domestic sanitary fittings is usually taken from a hot water tank or cylinder. The source of heat is usually in the form of a gas-fired, oil-fired or solid-fuel boiler; alternatives are a back boiler to an open fire or an electric immersion heater fixed into the hot water storage tank. When a quantity of hot water is drawn from the storage tank it is immediately replaced by cold water from the cold water storage cistern. Two main systems are used to heat the water in the tank – these are called the direct and indirect systems. For any hot water system copper or steel pipes are generally used. These materials are satisfactory if used independently throughout the system, but they must not be mixed or interchanged within a system. This is due to the electrolytic attraction that dissimilar metals have in the presence of water, which eventually leads to corrosion. Copper and zinc are particularly renowned for this problem, as zinc is the protective plating given to steel pipes under the name of galvanizing.

DIRECT HOT WATER SYSTEM

This is the simplest and cheapest system; the cold water flows through the water jacket in the boiler where its temperature is raised and convection currents are induced which causes the water to rise and circulate. The hot water leaving the boiler is replaced by colder water descending from the hot water cylinder or tank by gravity thus setting up the circulation. The hot water supply is drawn off from the top of the cylinder by a horizontal pipe at least 450 mm long to prevent 'one pipe' circulation being set up in the vent or expansion pipe. This pipe is run vertically from the hot water distribution pipe to a discharge position over the cold water storage cistern (see Fig. 10.1.4).

The direct system is not suitable for supplying a central heating circuit or for hard water areas because the pipes and cylinders will become furred with lime deposits. This precipitation of lime occurs when hard water is heated to temperatures of between 50 and 70 °C, which is the ideal temperature range for domestic hot water supply.

INDIRECT HOT WATER SYSTEM

This system is designed to overcome the problems of furring which occurs with the direct hot water system. The basic difference is in the cylinder design which now becomes a heat exchanger. The cylinder contains a coil or annulus which is connected to the flow and return pipes from the boiler. A transfer of heat takes place within the cylinder and therefore, after the initial precipitation of lime within the primary circuit and boiler, there is no further furring since fresh cold water is not being constantly introduced into the boiler circuit.

The supply circuit from the cylinder follows the same pattern as the direct hot water system, but a separate feed and expansion system is required for the boiler and primary circuit for initial supply, also for any necessary topping up due to evaporation. The feed cistern is similar to a cold water storage cistern but of a much smaller capacity. The water levels in the two cisterns should be equal so that equal pressures act on the indirect cylinder.

A gravity heating circuit can be taken from the boiler, its distribution being governed by the boiler capacity (see Fig. 10.1.4). Alternatively a small–bore forced system of central heating may be installed.

Hot water cylinders and tanks

Galvanized steel tanks, which are rectangular, can be used for any hot water system with compatible pipework. They are useful where space is restricted and the required storage capacity is between 95 and 115 litres as specified in BS 417. Cylinders of galvanized mild steel are produced to the same British Standard in capacities of 71 to 441 litres. Copper cylinders are produced to the recommendations of BS 1566. The standard recommends sizes, capacities and positions for screwed holes for pipe connections.

Note: In hard water areas primary circuit to be 35 mm

Direct hot water system

Indirect hot water system

Note: Pipe sizes shown are for standard copper outside diameters

Figure 10.1.4 Hot water system

To overcome the disadvantage of the extra pipework involved when using an indirect cylinder, a single feed indirect or 'Primatic' cylinder can be used. This form of cylinder is entirely self-contained and is installed in the same manner as a direct cylinder but functions as an indirect cylinder. It works on the principle of the buoyancy of air which is used to form seals between the primary and secondary water systems. When the system is first filled with water the cylinder commences to fill and fully charges the primary circuit to the boiler with water. When the cylinder water capacity has been reached two air seals will have formed, the first being in the upper chamber of the primatic unit and the second in the air vent pipe. These volumes of air are used to separate the primary and secondary water. When the water is heated in the primary system expansion displaces some of the air in the upper chamber to the lower chamber, this is a reciprocating action; the seals transfer from chamber to chamber as the temperature rises and falls.

Any excess air in the primary system is vented into the secondary system, which will also automatically replenish the primary system should this be necessary. As with indirect systems careful control over the heat output of the boiler is advisable to prevent boiling and consequent furring of the pipework. Typical examples of cylinders are shown in Fig. 10.1.5.

Faults in hot water systems

Unless a hot water system is correctly designed and installed a number of faults may occur such as air locks and noises. Air locks are pockets of trapped air in the system which will stop or slow down the circulation. Suspended air in the water will be released when the water is heated, and rise to the highest point. In a good installation the pipes are designed to rise 25 mm in 3,000 mm towards the vent where the air is released through the vent pipe. The most common positions for air locks are sharp bends and the upper rail of a towel rail; the only cure for the latter position is for the towel rail to be vented.

Noises from the hot water system usually indicate a blocked pipe caused by excessive furring or corrosion. The noise is caused by the imprisoned expanded water and the faulty pipe must be descaled or removed, or an explosion may occur.

Inadequate bracketing of pipes is another common cause of plumbing noise, particularly in the rising main where pressures are higher than elsewhere. This is known as 'water hammer'.

Mains supply hot water

These have become a popular alternative to traditional installations. They are very economic in pipework, space requirements and installation time as conventional expansion facilities are not required. The system is mains fed and sealed, with expansion accommodated in a specially designed vessel containing an air 'cushion'. Safety facilities are essential and include both pressure and thermal relief valves. Figure 10.1.6 shows an outline indicating the principles of operation.

114 litre indirect cylinder

135 litre single feed or 'Primatic' cylinder

Figure 10.1.5 Typical hot water cylinders

Figure 10.1.6 Schematic of mains fed hot water supply

SANITARY FITTINGS AND PIPEWORK

Sanitary fittings or appliances can be considered under two headings:

- **Soil fitments**   Those which are used to remove soil water and human excreta such as water closets and urinals.
- **Waste water fitments**   Those which are used to remove the waste water from washing and the preparation of food including appliances such as wash basins, baths, showers and sinks.

All sanitary appliances should be made from impervious materials, be quiet in operation, easy to clean and be of a convenient shape fixed at a suitable height. A number of materials are available for most domestic sanitary fittings.

Vitreous china   This is a white clay body which is vitried and permanently fused with a vitreous glazed surface when fired at a very high temperature, generally to the recommendations of BS 3402. Appliances made from this material are non-corrosive, hygienic and easily cleaned with a mild detergent or soap solution.

Glazed fireclay   This consists of a porous ceramic body glazed in a similar manner to vitreous china; it is exceptionally strong and resistant to impact damage but will allow water penetration of the body if the protective glazing is damaged. Like vitreous china, these appliances are non-corrosive, hygienic and easily cleaned.

Vitreous enamel   This is a form of glass which can be melted and used to give a glazed protective coating over a steel or cast iron base. Used mainly for baths, sinks and draining boards, it produces a fitment which is lighter than those produced from a ceramic material, is hygienic, easy to clean and has a long life. The finish, however, can be chipped and is subject to staining, especially from copper compounds from hot water systems.

Plastic materials   Acrylic plastics, glass-reinforced polyester resins and polypropylene sanitary fittings made from these plastics require no protective coatings, are very strong, light in weight, chip resistant but generally cost more than ceramic or vitreous enamel products. Care must be taken with fitments made of acrylic plastics since these become soft when heated, therefore they should be

474

used for cold water fitments or have thermostatically controlled mixing taps. Plastic appliances can be easily cleaned using warm soapy water and any dullness can be restored by polishing with ordinary domestic polishes.

Stainless steel   This is made from steel containing approximately 18% chromium and 8% nickel which gives the metal a natural resistance to corrosion. Stainless steel appliances are very durable and relatively light in weight; for domestic situations the main application is for sinks and draining boards. Two finishes are available: polished or 'mirror' finish and 'satin' finish; the latter has a greater resistance to scratching.

General considerations   The factors to be considered when selecting or specifying sanitary fitments can be enumerated thus:

1. Cost: outlay, fixing and maintenance.
2. Hygiene: inherent and ease of cleaning.
3. Appearance: size, colour and shape.
4. Function: suitability, speed of operation and reliability.
5. Weight: support required from wall and/or floor.
6. Design: ease with which it can be included into the general services installation.

Building Regulation G1 establishes requirements for sanitary conveniences, i.e. water closets, urinals and wash basins, while Regulation G2 extends coverage to bathrooms containing bath or shower installations. The quantity of WCs and sanitary appliances in various locations is detailed in BS 6465: *Sanitary installations*. This Standard is in two parts, Part 1 the Code of practice for provision and installation and Part 2 relates to spacial layout and design for sanitary accommodation.

The subject of water conservation with respect to use of sanitary fitments is very topical, with manufacturers endeavouring to adapt their products and develop new ideas in line with conservation and environmental issues. Building Research Establishment, Information Paper ref. IP 8/97, Parts 1 and 2, provides further reading on this topic with application to the design of low–flush WCs.

WATER CLOSETS

Most water closets are made from a ceramic base to the requirements of BS 5503 with a horizontal outlet. The complete water closet arrangement consists of the pan, seat, flush pipe and flushing cistern. The cistern can be fixed as a high level, low level or closed coupled; the latter arrangement dispenses with the need for a flush pipe. A typical arrangement is shown in Fig. 10.2.1. The BS 5503 water closet is termed a wash down type and relies on the flush of water to wash the contents of the bowl round the trap and into the soil-pipe. An alternative form is the siphonic water closet which is more efficient and quieter in operation but has a greater risk of blockage if misused. Two types are produced – the single trap and the double trap.

Plastic WC seat
to BS 1254

Metal or plastic
flush pipe

One-piece rubber
or plastic connector

Flushing
rim

Flushing
cistern
to BS 1125
or BS 7358

Low level
1.060 maximum

High level
1.520 or 1.670

150 × 110
min.

390

50 min.

Horizontal
outlet used
in conjunction
with converters
to form P or S
trap.

Distance from wall governed
by flushing system

90 mm dia. outlet

BS 5503 ceramic washdown WC pan

After flush
chamber

Flushing
Rim

After flushing
chamber

813

406

406

711

711

51

Typical siphonic WC pans

Warning
pipe

Lever

Pull
chain

Flush pipe

Cast iron
bell

Well bottom

Piston
chamber

Siphon bend

Warning
pipe

Piston
and flap

Flush pipe

Bell type cistern

Piston type cistern

Figure 10.2.1 WC pans and cisterns

The single-trap siphonic water closet has a restricted outlet which serves to retard the flow, when flushed, so that the bore of the outlet connected to the bowl becomes full and sets up a siphonic flushing action, completely emptying the contents of the bowl (see Fig. 10.2.1). With the double-trap siphonic pan the air is drawn from the pocket between the two traps; when the flushing operation is started this causes atmospheric pressure to expel the complete contents of the bowl through both traps into the soil pipe (see Fig. 10.2.1).

The pan should be fixed to the floor with brass screws and bedded on a suitable compressible material; the connection to the soil-pipe socket can be made with cement mortar or preferably a mastic to allow for any differential movement between the fitment and the structure. Connections to PVC soil pipes are usually made with compression rings. The flush pipe is invariably connected to the pan with a special plastic or rubber one-piece connector.

Flushing cisterns together with the flush pipes are usually constructed to the recommendations of BS 1125 and BS 7358 and can be made from enamelled cast iron, enamelled pressed steel, ceramic ware or of plastic materials. Two basic types have been produced, namely the bell or cone and the piston. The former is now obsolete, but may still be found in service. It is activated by pulling a chain which raises and lowers the bell or cone and in so doing raises the water level above the open end of the flush pipe thus setting up a siphonic action. These cisterns are efficient and durable but are noisy in operation (see Fig. 10.2.1). The piston type cistern is the one in general use and is activated by a lever or button. When activated the disc or flap valve piston is raised and with it the water level which commences the siphonage (see Fig. 10.2.1). The level of the water in the cistern is controlled by a float valve and an overflow or warning pipe of a larger diameter than the inlet is fitted to discharge so that it gives a visual warning, usually in an external position. The capacity of the cistern will be determined by local water board requirements, the most common being 6, 7.5 and 9 litres.

There is a wide range of designs, colours and patterns available for water closet suites but all can be classified as one of the types described above.

FLUSHING VALVES

These have been a popular marine plumbing installation for many years. In compact lavatory compartments they have proved effective in restricted space and are economic with the use of limited water resources. They are becoming a creditable alternative for use in buildings, particularly where usage is frequent as they can flush at any time without delay. They comprise a large equilibrium valve which functions on the principle of water displacement and pressure equalization limiting the flow. Current research and recommendations from the Water Regulations Advisory Committee (WRAC) to the Government are likely to succeed in establishing these valves in The Water Supply Byelaws. See BRE IP 8/97 Part 2 for more detail.

WASH BASINS

Wash basins for domestic work are usually made from a ceramic material but metal basins complying with BS 1329 are also available. A wide variety of shapes, sizes, types and colours are available, the choice usually being one of personal preference. BS 1188 and BS 5506 provide recommendations for ceramic wash basins and

Figure 10.2.2   Basins, traps and wastes

pedestals and specify two basic sizes, 635 × 457 and 559 × 406. Small wall-hung hand-rinse basins are also available for use in limited space. These are produced to BS 6731. All these basins are a one-piece fitment having an integral overflow, separate waste outlet and generally pillar taps (see Fig. 10.2.2).

Wash basins can be supported on wall-mounted cantilever brackets, leg supports or pedestals. The pedestals are made from identical material to the wash basin and are recessed at the back to receive the supply pipes to the taps and the waste pipe from the bowl. Although pedestals are designed to fully support the wash basin most manufacturers recommend that small wall-mounted safety brackets are also used.

BATHS AND SHOWERS

Baths are available in a wide variety of designs and colours made either from porcelain-enamelled cast iron, vitreous-enamelled sheet steal, 8 mm cast acrylic sheet or 3 mm cast acrylic sheet reinforced with a polyester resin/glass fibre laminate. Most bath designs today are rectangular in plan and made as flat bottomed as practicable with just sufficient fall to allow for gravity emptying and resealing of the trap. The British Standards for the materials quoted above recommend a co-ordinating plan size of 1.700×700 with a height within the range of $n \times 50$ mm, where n is any natural number including unity. Baths are supplied with holes for pillar taps or mixer fittings and for the waste outlets. Options include handgrips, built-in soap and sponge recesses and overflow outlets. It is advisable always to specify overflow outlets as a precautionary measure to limit the water level and to minimize splashing; most overflow pipes are designed to connect with the waste trap beneath the bath (see Fig. 10.2.3). Support for baths is usually by adjustable feet for cast iron and steel and by a strong cradle for the acrylic baths. Panels of enamelled hardboard or moulded high-impact polystyrene or glass fibre are available for enclosing the bath. These panels can be fixed by using stainless steel or aluminium angles or direct to a timber stud framework.

Shower sprays can be used in conjunction with a bath by fitting a rigid plastic shower screen or flexible curtain to one end of the bath. A separate shower fitment, however, is considered preferable. Such fitments require less space than the conventional bath, use less hot water and are considered to be more hygienic since the used water is being continuously discharged. A shower fitment consists of the shower tray with a waste outlet, the impervious cubicle and a door or curtain (see Fig. 10.2.3). Materials available are similar to those described for baths. The spray outlet is normally fixed to the wall and is connected to a mixing valve so that the water temperature can be controlled.

SINKS

Sinks are used mainly for the preparation of food, washing of dishes and clothes, and are usually positioned at the drinking water supply outlet. Their general design follows that described for basins except that they are larger in area and deeper. Any material considered suitable for sanitary appliance construction can be used. Designs range from the simple belfast sink with detachable draining boards of metal, plastic or timber to the combination units consisting of integral draining

Co-ordinating width = 700; work size = 697
Co-ordinating sizes can be varied within range of
$n \times 100$ mm, where n = any natural number including unity

BS 1189 (cast iron), BS 4305 (acrylic) and BS 1390 (vitreous enamelled steel) baths. Also, BS EN 60335–2–60 whirlpool baths.

Alternative overflow connections

Typical sizes
$600 \times 600 \times 175$ deep
$760 \times 760 \times 175$ deep
$900 \times 900 \times 175$ deep
Also available in enamelled steel and acrylic to BS 7015

Typical fireclay shower tray

Figure 10.2.3 Baths and shower trays

boards and twin bowls. Support can be wall-mounted cantilever brackets, framed legs or a purpose made cupboard unit; typical details are shown in Fig. 10.2.4.

The layout of domestic sanitary appliances is governed by size of fitments, personal preference, pipework system being used and the space available. Building Regulations G1 and G2 detail requirements and minimum facilities to be made

Weir overflow

Outlet for 40 mm waste and trap connection

BS 1206 fireclay sinks
Belfast pattern – wide range of sizes
from 457 × 380 × 200 deep up to
1.219 × 610 × 305 deep. Sinks
supported on cantilever brackets,
legs and bearers or on brick
dwarf walls. Fixing heights 850
to 920 to top of sink

Edge
lipping

Flutes

Clip-on draining boards
Reversible and available in
hardwood, stainless steel,
aluminium alloy and vitreous
enamel in range of sizes
from 560 × 406 to 762 × 457

Brackets
hook over
sink

Hole for
pillar tap

Overflow slot

Stainless steel sinks
Generally made to the
requirements of BS 1244
available with single or
double drainers in a
range of sizes from 1.070 ×
460 to 1.600 × 530. Sink
tops supported on
cantilever brackets

180 or 200
deep bowl

Single
drainer

Mixer taps

Sink units
Available in same plan sizes
as BS 1244 sinks to a general
work height of 910 mm. Sink
top can be of stainless steel
or vitreous enamel with a wide
choice of cabinet designs
and finishes

Toe
recess

Sliding
cupboard doors

Note: Waste fittings to BS EN 411

Figure 10.2.4 Sinks and draining boards

available in a dwelling and the need for sanitary accommodation to be separated from kitchen areas.

PIPEWORK

Building Regulations, Approved Document H1, Sanitary Pipework and Drainage, sets out in detail the recommendations for soil pipes, waste pipes and ventilating pipes. These regulations govern such things as minimum diameters of soil pipes, material requirements, provision of adequate water seals by means of an integral trap or non-integral trap, the positioning of soil pipes on the inside of a building, overflow pipework and ventilating pipes. The only pipework which is permissible on the outside of the external wall is any waste pipe from a waste appliance situated at ground floor level providing such a pipe discharges into a suitable trap with a grating and the discharge is above the level of the water but below the level of the grating.

PIPEWORK SYSTEMS

Three basic pipework systems have been used for domestic work, namely:

- One-pipe system.
- Two-pipe system.
- Single-stack system.

Whichever system is adopted the functions of quick, reliable and quiet removal of the discharges to the drains remains constant.

One-pipe system

This consists of a single discharge pipe which conveys both soil and waste water directly to the drain. To ensure that water seals in the traps are not broken, deep seals of 75 mm for waste pipes up to 65 mm diameter and 50 mm for pipes over 75 mm diameter are required. To allow for unrestricted layout of appliances most branch pipes will require an anti-siphon arrangement (see Fig. 10.2.5). The advantage of this system is the flexibility of appliance layout; the main disadvantage is cost; and generally the one-pipe system has been superseded by the more restricted but economic single-stack system described below.

Two-pipe system

As its name implies, this system consists of two discharge pipes, one conveys soil discharges and the other all the waste discharges. It is a simple, reliable and costly system but has the advantages of complete flexibility in appliance layout and deep seal traps are not usually required. Like the one-pipe system, it has been largely superseded by the single-stack system. A comparison of the one- and two-pipe systems is shown in Fig. 10.2.5.

Figure 10.2.5 Comparison of one-pipe and two-pipe systems

Two-pipe system

WC

Waste pipes

Soil and vent pipe

WC

50 mm min. seal

Back inlet gully

Sink

Waste pipe

Direct connection

Waste and vent pipe

Bath

Except WCs, all traps to have 40 mm min. seal

Basin

Vent pipes terminated above roof level with balloon cages

Simple but more expensive method

Economic if appliances grouped around soil vent pipe

Layout of appliances unrestricted

One-pipe system

25 mm anti-siphon or ventilating pipe to permit entry of air to equalize pressures

Bath

40 mm waste pipe

Soil vent pipe

WC

50 mm min. seal

Direct connection

25 mm asp

WC

32 mm waste pipe

Sink

Back inlet gully

40 mm waste pipe

Basin

Except WCs, all traps to have 75 mm min. seal

Figure 10.2.6 Single-stack system

Single-stack system

This system was developed by the Building Research Establishment and is fully described in their Digest No. 249. It is also fully detailed in BS 5572 and is now adapted into the Building Regulations, Approved Document H1. It is a simplification of the one-pipe system by using deep seal traps, relying on venting by the discharge pipe and placing certain restrictions on basin waste pipes which have a higher risk to self-siphonage than other appliances. A diagrammatic layout is shown on Fig. 10.2.6.

Materials which can be used for domestic stack pipework include galvanized steel prefabricated stack units (BS 3868), cast iron (BS 416) and uPVC (BS 4514). The latter is standard contemporary practice, being light and easy to install with simple push fit or solvent weld joints. Branch waste pipes can be of polypropylene (BS 5254), thermoplastic (BS 5255) or copper (BS EN 1057).

Most manufacturers of soil pipes, ventilating pipes and fittings produce special components for various plumbing arrangements and appliance layouts. These fittings have the water closet socket connections, bosses for branch waste connections and access plates for cleaning and maintenance arranged as one prefabricated assembly to ease site work and ensure reliable and efficient connections to the discharge or soil pipe.

DRAINAGE

Drainage is a system of pipework, usually installed below ground level, to convey the discharge from sanitary fittings, rainwater gutters and downpipes to a suitable disposal installation. The usual method of disposal is to connect the pipework to the public sewer which will convey the discharges to a local authority sewage treatment plant. Alternatives are a small self contained treatment plant on site or a cesspool; the latter is a collection tank to hold the discharge until it can be collected in a special tanker lorry and taken to the local authority sewage treatment installation for disposal.

PRINCIPLES OF GOOD DRAINAGE

1. Materials should have adequate strength and durability.
2. Diameter of drain to be as small as practicable: for soil drains the minimum diameter allowed is 100 mm and for surface water the minimum diameter is 75 mm.
3. Every part of a drain should be accessible for the purposes of inspection and cleansing.
4. Drains should be laid in straight runs as far as possible.
5. Drains must be laid to a gradient which will render them efficient. The fall or gradient should be calculated according to the rate of flow, velocity required and the diameter of the drain. Individual domestic buildings have an irregular flow and tables to accommodate this are provided in the Building Regulations, Approved Document H1. Alternatively, Maguire's rule can be used to give a gradient with reasonable velocity. This well-established rule of thumb calculates the gradient by dividing the pipe diameter (mm) by 2.5. Thus a 100 mm drain can have a fall of 1 in 40, a 150 mm drain 1 in 60, and so on. Velocity of flow should be at least 0.75 m/s to ensure self-cleansing with a drain capacity not more than 0.75 proportional depth to ensure clear flow conditions. For more detailed drainage calculations, see *Building Services, Technology and Design* by R. Greeno (Longman and CIOB co-publishers).

6. Every drain inlet should be trapped to prevent the entry of foul air into the building, the minimum seal required is 50 mm. The trap seal is provided in many cases by the sanitary fitting itself; rainwater drains need not be trapped unless they connect with a soil drain or sewer.
7. Inspection chambers, manholes, rodding eyes or access fittings should be placed at changes of direction and gradient if these changes would prevent the drain from being readily cleansed.
8. Inspection chambers must also be placed at a junction, unless each run can be cleared from an access point.
9. Junctions between drains must be arranged so that the incoming drain joints at an oblique angle in the direction of the main flow.
10. Avoid drains under buildings if possible; if unavoidable they must be protected to ensure watertightness and to prevent damage. The usual protection methods employed are:
    ■ Encase the drain with 100 mm (minimum) of granular filling.
    ■ Use cast iron pipes under the building.
11. Drains which are within 1 m of the foundations to the walls of buildings and below the foundation level must be backfilled with concrete up to the level of the underside of the foundations. Drains more than 1 m from the foundations are backfilled with concrete to a depth equal to the distance of the trench from the foundation less 150 mm.
12. Where possible the minimum invert level of a drain should be 450 mm to avoid damage by ground movement and 700 mm for traffic. The invert level is the lowest level of the bore of a drain.

DRAINAGE SCHEMES

The scheme or plan layout of drains will depend upon a number of factors:

■ Number of discharge points.
■ Relative positions of discharge points.
■ Drainage system and location of the local authority sewers.
■ Internal layout of sanitary fittings.
■ External positions of rainwater pipes.
■ Disposition of buildings.
■ Topography of the area to be served.

Drainage systems must be designed within the limits of the terrain, so that the discharges can flow by gravity from the point of origin to the point of discharge. The pipe sizes and gradients must be selected to provide sufficient capacity to accommodate maximum flows, while at the other extreme have adequate self-cleansing velocity at minimum flows to prevent debris accumulating. Economic and construction factors control the depth to which drains can be laid. Therefore, in flat and opposingly inclined areas it may be necessary to provide pumping stations to raise the drainage discharge to higher level sewers.

There are three drainage systems used by local authorities in this country and the method employed by any particular authority will determine the basic scheme to be used for the drain runs from individual premises.

COMBINED SYSTEM

All the drains discharge into a common or combined sewer. It is a simple and economic method since there is no duplication of drains. This method has the advantages of easy maintenance, all drains are flushed when it rains and it is impossible to connect to the wrong sewer. The main disadvantage is that all the discharges must pass through the sewage treatment installation, which could be costly and prove to be difficult with periods of heavy rain.

TOTALLY SEPARATE SYSTEM

The most common method employed by local authorities; two sewers are used in this method. One sewer receives the surface water discharge and conveys this direct to a suitable outfall such as a river where it is discharged without treatment. The second sewer receives all the soil or foul discharge from baths, basins, sinks, showers and toilets; this is then conveyed to the sewage treatment installation. More drains are required and it is often necessary to cross drains one over the other. There is a risk of connection to the wrong sewer and the soil drains are not flushed during heavy rain, but the savings on the treatment of a smaller volume of discharge leads to an overall economy.

PARTIALLY SEPARATE SYSTEM

This is a compromise of the other two systems and is favoured by some local authorities because of its flexibility. Two sewers are used, one to carry surface water only and the other to act as a combined sewer. The amount of surface water to be discharged into the combined sewer can be adjusted according to the capacity of the sewage treatment installation.

Soakaways, which are pits below ground level designed to receive surface water and allow it to percolate into the soil, are sometimes used to lessen the load on the surface water sewers. Typical examples of the three drainage systems are shown in Fig. 10.3.1.

PRIVATE SEWERS

A sewer can be defined as a means of conveying waste, soil or rainwater below the ground that has been collected from the drains and conveying it to the final disposal point. If the sewer is owned and maintained by the local authority it is generally called a public sewer whereas one owned by a single person or a group of people and maintained by them can be classed as a private sewer.

Figure 10.3.1 Drainage systems

When planning the connection of houses to the main or public sewer one method is to consider each dwelling in isolation but important economies in design can be achieved by the use of a private sewer. A number of houses are connected to the single sewer which in turn is connected to the public sewer. Depending upon the number of houses connected to the private sewer and the distance from the public sewer the following savings are possible:

1. Total length of drain required.
2. Number of connections to public sewer.
3. Amount of openings in the roads.
4. Number of inspection chambers.

A comparative example is shown in Fig. 10.3.2.

DRAINAGE MATERIALS

Drain pipes are considered as either rigid or flexible according to the material used in their manufacture. Clay is a major material used for rigid drain pipes in domestic work, with cast iron as an expensive alternative. Flexible drain pipes are produced from unplasticized PVC.

Clay pipes

This is the traditional material used for domestic drainage, with current manufacturing standards in accordance with BS 65 and BS EN 295. These pipes were treated with a glazing of common salt, borax or boric acid during the firing process to render them impervious. Today's high standards of manufacture combined with quality dense materials no longer justify this process. Several qualities of pipes are produced ranging from standard pipes for general use, surface water pipes and pipes of extra strength to be used where heavy loadings are likely to be encountered. The type and quality of pipes are marked on the barrel so that they can be identified after firing.

Clay pipes are produced in a range of diameters from 100 to 300 mm nominal bore for general building work. Lengths vary between manufacturers from 1,300 to 1,600 mm for plain end pipes, with spigot and socketed pipes produced in shorter lengths down to 600 mm. They can be obtained with sockets and spigots prepared for rigid or flexible jointing. Rigid jointing is rarely used now, except perhaps for small repairs or alterations to an existing drain as it is difficult to make a sound joint in wet trenches, plus the inability of the system to absorb movement without fracturing. Most pipes are supplied plain ended for use with push fit polypropylene sleeve couplings. A wide variety of fittings for use with clay pipes are manufactured to give flexibility when planning drainage layouts and means of access. Typical examples of clay pipes, joints and fittings are shown in Figs 10.3.3 and 10.3.4.

Clay pipes are resistant to attack by a wide range of acids and alkalis; they are therefore suitable for all forms of domestic drainage.

Individual drain and sewer connections

Note: Generally 20 houses can be connected to a 100 mm dia.
private sewer at a gradient of 1:70 and 100 houses
can be connected to a 150 mm dia. private sewer
laid to a fall of 1:150

Figure 10.3.2   Example of a private sewer arrangement

Cast iron pipes

These pipes are generally only considered for domestic drainage in special
circumstances such as sites with unstable ground, drains with shallow inverts and
drains which pass under buildings. Like clay pipes, cast iron pipes are made with a

Typical BS 65 clay drain pipe

Rigid joint
(Not recommended in AD, H)

Flexible 'O' ring joint

Flexible coupling joint

Flexible ring seal socket joint

Pitch fibre pipe joints
(Not recommended in AD, H)

Figure 10.3.3   Pipes and joints

45° oblique junction   **Eighth bend**

$\frac{1}{4}$ $\frac{1}{6}$ $\frac{1}{12}$ $\frac{1}{16}$ $\frac{1}{24}$ and $\frac{1}{32}$ bends also available

$\frac{1}{2}$ and $\frac{3}{4}$ section channel junctions and bends similar

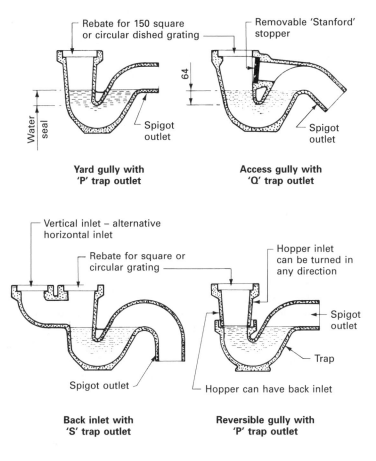

Yard gully with 'P' trap outlet   **Access gully with 'Q' trap outlet**

Back inlet with 'S' trap outlet   **Reversible gully with 'P' trap outlet**

All gullies can be supplied with 'P', 'Q', or 'S' trap outlets

Figure 10.3.4   Clay drain pipe fittings and gullies

spigot and socket for rigid or flexible joints. The rigid joint is made with a tarred gaskin and caulked lead whereas the flexible joint has a sealing strip in the socket allowing a 5° deflection. Lengths, diameters and fittings available are similar to those produced for clay pipes but to the recommendations of BS 437 and BS EN 598.

Cast iron pipes are given a protective coating of a hot tar composition or a cold solution of a naphtha and bitumen composition. This coating gives the pipes good protection against corrosion and reasonable durability in average ground conditions.

Pitch fibre pipes

Pitch fibre pipes and fittings preceded uPVC as a flexible drainage material. They are rarely used now, but will be found in many existing systems. They are made from preformed felted wood cellulose fibres thoroughly impregnated under vacuum and pressure, with at least 65% by weight of coal tar pitch or bituminous compounds. They have been used for all forms of domestic drainage and because of the smooth bore with its high flow capacity, could be laid to lower gradients than most other materials of their day. Sizes range from 50 to 225 mm nominal bore, with lengths generally 2,400 and 3,000 mm.

The original joints had a machined 2° taper on the ends of the pipe which made a drive fit to machined pitch fibre couplings. These joints are watertight but do not readily accommodate axial or telescopic movement. The snap joint used a rubber 'D' ring in conjunction with a polypropylene coupling giving a flexible joint. Both methods are shown in Fig. 10.3.3.

Unplasticized PVC pipes

These pipes and fittings are made from polyvinyl chloride plus additives which are needed to facilitate the manufacture of the polymer and produce a sound, durable pipe. BS 3506 gives the requirements for pipes intended for industrial purposes where BS 4660 covers the pipes and fittings for domestic use. Standard outside diameters are 110 and 160 mm (100 and 150 mm nominal bore) with non-standard outside diameters of 82, 200, 250 and 315 mm being available from most manufacturers. Long lengths of 3 m and 6 m can be easily cut by hacksaw. BS 5481 is also available as a specification for uPVC sewer pipes in nominal diameters of 200 to 630 mm. The pipes are obtainable with socket joints for either a solvent welded joint or a ring seal joint (see Fig. 10.3.3). Like pitch fibre pipes, uPVC pipes have a smooth bore, are light and easy to handle – long lengths reducing to a minimum the number of joints required; they can be jointed and laid in all weathers.

DRAIN LAYING

Domestic drains are laid in trenches which are excavated and if necessary timbered in a similar manner to that described for foundations; the main difference is that drain trenches are excavated to the required fall or gradient. It is good practice to programme the work to enable the activities of excavation, drain laying and backfilling to be carried out in quick succession so that the excavations remain open for the shortest possible time.

The technique used in the laying and bedding of drains will depend upon two factors:

■ Material – rigid or flexible.
■ Joint – rigid or flexible.

Approved Document H recommends drains to be of sufficient strength, durability and so jointed that the drain remains watertight under all working conditions, including any differential movement between the pipe and ground.

Many examples of traditional bedding and haunching in concrete will be found in existing drains, but this is no longer practised unless the drain requires a complete surround of concrete for total protection. The cost is prohibitive, the quality of concrete difficult to control in wet trenches and the system is prone to failure in ground movement and settlement. Both clay and uPVC drains can accommodate axial flexibility and extensibility by combining flexible jointing with a granular flexible bedding medium. Several examples of bedding specifications are provided in the Building Regulations, Approved Document H and BS 8301, the Code of practice for building drainage. Two popular applications are shown in Fig. 10.3.5.

The selected material required for granular bedding and for tamping around pipes laid on a jointed concrete base must be of the correct quality. Pipes depend to a large extent upon the support bedding for their strength and must therefore be uniformly supported on all sides by a material which can be hard compacted. Generally a non-cohesive granular material with a particle size of 5–20 mm is suitable and if not present on site it will have to be 'imported'.

Pipes with socket joints are laid from the bottom of the drain run with the socket end laid against the flow, each pipe being aligned and laid to the correct fall. The collar of the socket is laid in a prepared 'hollow' in the bedding and the bore is centralized. In the case of a rigid joint a tarred gaskin is used which also forms the seal, whereas the mechanical or flexible joints are self aligning. Most flexible joints require a special lubricant to ease the jointing process and those which use a coupling can be laid in any direction.

Means of access

Drains require access for testing, maintenance and clearance of blockages. Four possibilities exist, these are:

■ Rodding eye
■ Shallow access fitting
■ Inspection chamber
■ Manhole

Figure 10.3.5   Typical pipe bedding details

Figure 10.3.6 Drains – means of access up to 1 m deep

Rodding eyes These generally locate at the head of a drain and are effectively a swept extension of the drain to surface level, with a sealed access plate. With local authority approval they can replace a more expensive inspection chamber to provide means for clearance in one direction only.

Shallow access fittings These provide vertical access to both directions of a drain run which is not over 600 mm deep.

Inspection chambers These are an enlarged version of the shallow access fitting, used to invert depths of up to 1 m. They provide for limited access and contain facilities for a few junctions and branch connections. Materials include plastics, precast concrete and traditional brick masonry. Figure 10.3.6 shows examples of means of drain access, with the plastic inspection chamber containing optional adaptors which can be cut to suit 100 or 150 mm pipes at various approach angles.

The depth of these units is cut to suit drain trench levels and a cast iron or aluminium frame and cover is fitted.

Manholes   These are inspection chambers over 1 m to invert. They are a compartment containing half or three-quarter section round channels to enable the flow to be observed and to provide a drain access point for cleansing and testing. Both inspection chambers and manholes are positioned to comply with the access recommendations of Approved Document H (see Fig. 10.3.1).

Simple domestic drainage is normally only concerned with shallow manholes up to an invert depth of 1,800 mm. The internal sizing is governed by the depth to invert, the number of branch drains, the diameter of branch drains and the space required for a person to work within the manhole. A general guide to the internal sizing of inspection chambers and manholes is given in Approved Document H, Table 9.

Manholes can be constructed of brickwork or of rectangular or circular precast concrete units (see Fig. 10.3.7). The access covers used in domestic work are generally of cast iron or pressed steel and light duty as defined in BS EN 124. They have a single seal which should be bedded in grease to form an air-tight joint; double seal covers would be required if the access was situated inside the building. Concrete access covers are available for use with surface water manholes.

Ventilation of drains

To prevent foul air from soil and combined drains escaping and causing a nuisance, all drains should be vented by a flow of air. A ventilating pipe should be provided at or near the head of each main drain and any branch drain exceeding 10.000 in length. The ventilating pipe can be a separate pipe or the soil discharge stack pipe can be carried upwards to act as a ventilating discharge stack or soil vent pipe. Ventilating pipes should be open to the outside air and carried up for at least 900 mm above the head of any window opening within a horizontal distance of 3.000 from the ventilating pipe which should be finished with a cage or cover which does not restrict the flow of air.

Conventional venting through the stack is not always necessary. An air admittance valve can be fitted in up to four consecutive dwellings of no more than three storeys, if the fifth dwelling is conventionally vented. Systems with air admittance valves have the following advantages:

- Ventilating stack can terminate inside the building (above highest spill over level), typically in the roof space.
- Greater design flexibility.
- Adaptable to plastic or metal pipework.
- Visually inobtrusive as there is no projecting stack pipe.

The principles of application and operation are shown in Fig. 10.3.8. A discharge of water in the stack creates a slight negative pressure, sufficient to open the valve

Brick-levelling course

600 × 450 light duty cast iron cover and frame to BS EN 124

100 mm precast concrete slab

One brick wall in BS 3921 class B bricks bedded in cm. mt. (1:3) and laid to English bond

Main drain channel

Smooth cement (1:1) finish

1:12

Branch drain

Mass concrete benching

150 mm thick mass concrete (1:3:6) base

Shallow brick inspection chamber

Frame bedded in cm. mt. (1:3)

600 × 450 light duty cast iron cover and frame to BS EN 124

150 mm concrete encasing in wet soils

Precast concrete cover slab

600 min. diameter

Precast concrete shaft rings circular or rectangular in plan

Main drain channel

150 to 600

Precast concrete base with channels and benching cast in as required

1:12

Branch drain

Precast concrete inspection chamber (BS 5911: Part 2)

Figure 10.3.7 Typical shallow inspection chambers

Figure 10.3.8   Air admittance valve

and admit air. After discharge a return to atmospheric pressure allows the spring to re-seal the unit to prevent foul air escaping. To satisfy building regulations, the device must have received design approval in accordance with the Board of Agrément.

Rainwater drainage

A rainwater drainage installation is required to collect the discharge from roofs and paved areas, and convey it to a suitable drainage system. Paved areas, such as garage forecourts or hardstands, are laid to fall so as to direct the rainwater into a yard gully which is connected to the surface water drainage system. A rainwater

For details of eaves
construction see
Figs 6.1.5 and 6.2.2

Half-round section
rainwater gutter

Gutter brackets
fixed to fascia
at 1.000 c/c

112° offset or
swanneck

Pipe clip or holderbat

Rainwater pipe

Cavity insulation
and insulated
blockwork inner
leaf if habitable
accommodation

Rainwater pipe

Sealed access cover

Raising piece

To surface
water drain

Rainwater shoe

Rainwater pipe

Sealed access cover

To combined
drain

Back inlet access gully

Figure 10.3.9 Rainwater pipework and drainage

installation for a roof consists of a collection channel called a 'gutter' which is
connected to vertical rainwater pipes. The rainwater pipe is terminated at its
lowest point by means of a rainwater shoe for discharge to a surface water drain
or a trapped gully if the discharge is to a combined drain (see Fig. 10.3.9). If a

separate system of drainage or soakaways is used it may be possible to connect the rainwater pipe direct to the drains, providing there is an alternative means of access for cleansing.

The materials available for domestic rainwater installations are asbestos, galvanized pressed steel, cast iron and uPVC. The usual materials for domestic work are cast iron and uPVC, the latter being the usual specification for new work.

Cast iron rainwater goods

Cast iron rainwater pipes, gutters and fittings are generally made to the requirements of BS 460 which specifies a half-round section gutter with a socket joint in diameters from 75 to 150 mm and an effective length of 1,800 mm. The gutter socket joint should be lapped in the direction of the flow and sealed with either putty or an approved sealing compound before being bolted together. The gutter is supported at 1,000 to 1,800 mm centres by means of mild steel gutter brackets screwed to the feet of rafters for an open eaves or to the fascia board with a closed eaves.

Cast iron rainwater pipes are also produced to a standard effective length of 1,800 mm with a socket joint which is sealed with putty, run lead or in many cases dry jointed – the pipe diameters range from 50 to 150 mm. The downpipes are fixed to the wall by means of pipe nails and spacers when the pipes are supplied with ears; or with split ring hinged holderbats when the pipes are supplied without ears cast on. A full range of fittings such as outlets, stopped ends, internal and external angles are available for cast iron half-round gutters, and, for the downpipes, fittings such as bends, offsets and rainwater heads are produced.

Unplasticized PVC rainwater goods

The advantages of uPVC rainwater goods over cast iron are:

1. Easier jointing, gutter bolts are not required and the joint is self sealing, generally by means of a butyl or similar strip.
2. Corrosion is eliminated.
3. Decoration is not required; several standard colours are available including brown, black, white and grey.
4. Breakages are reduced.
5. Better flow properties usually enable smaller sections and lower falls.

Half-round gutters are supplied in standard effective lengths up to 6 m with diameter ranging from 75 to 150 mm. Downpipes are supplied in two standard lengths of 2 and 4 m with diameters of 50, 63, 75 and 89 mm. The gutters, pipes and fittings are generally produced to the requirements of BS 4576. Typical details of domestic rainwater gutter and pipework are shown in Fig. 10.3.9.

Sizing of pipes and gutters

The sizing of the gutters and downpipes to effectively cater for the discharge from a roof will depend upon:

- The area of roof to be drained.
- Anticipated intensity of rainfall.
- Material of gutter and downpipe.
- Fall within gutter, usually in the range of 1/150 to 1/600.
- Number, size and position of outlets.

The requirements for Building Regulation H3 concerning rainwater drainage can be satisfied by using the design guide tables contained in Approved Document H which gives guidance to sizing gutters, downpipes and selection of suitable materials.

CONNECTIONS TO SEWERS

It is generally recommended that all connections to sewers shall be made so that the incoming drain or private sewer is joined to the main sewer obliquely in the direction of flow and that the connection will remain watertight and satisfactory under all working conditions. Normally sewer connections are made by the local authority or under their direction and supervision.

The method of connection will depend upon a number of factors:

1. Relative sizes of sewer and connecting drain or private sewer.
2. Relative invert levels.
3. Position of nearest inspection chamber on the sewer run.
4. Whether the sewer is existing or being laid concurrently with the drains or private sewers.
5. Whether stopped or joinder junctions have been built into the existing sewer.
6. The shortest and most practicable route.

If the public sewer is of a small diameter, less than 225 mm, the practical method is to remove two or three pipes and replace with new pipes and an oblique junction to receive the incoming drain. If three pipes are removed it is usually possible to 'spring in' two new pipes and the oblique junction and joint in the usual manner, but if only two pipes are removed a collar connection will be necessary (see Fig. 10.3.10).

If new connections have been anticipated stopped junctions or joinder junctions may have been included in the sewer design. A stopped junction has a disc temporarily secured in the socket of the branch arm whereas the binder has a cover cap as an integral part of the branch arm. In both cases careful removal of the disc or cap is essential to ensure that a clean undamaged socket is available to make the connection (see Fig. 10.3.10).

Removing 3 No. pipes and inserting oblique junction

Removing 2 No. pipes and inserting oblique junction

Joinder junction   **Stopped junction**

Figure 10.3.10 Connections to small-diameter sewers (traditional rigid-jointed spigot and socket clay pipes)

Connections to inspection chambers or manholes, whether new or existing, can take several forms depending mainly upon the differences in invert levels. If the invert levels of the sewer and incoming drain are similar the connection can be made in the conventional way using an oblique branch channel. Where there is a difference in invert levels the following can be considered:

- A ramp formed in the benching within the inspection chamber.
- A backdrop manhole or inspection chamber.
- Increase in the gradient of the branch drain.

The maximum difference between invert levels which can be successfully overcome by the use of a ramp is a matter of conjecture. The old CP 2005 entitled 'Sewerage', now replaced by BS 8005: Part 1 and BS EN 752, gave a maximum difference of 1.800 m, whereas BS 8301, covering building drainage, gives 1.000 for external ramps at 45°. The generally accepted limit of invert level difference for the use of internal ramps is 700 mm, which approximates to the figure quoted in the first edition of CP 301, since superseded by BS 8301. Typical constructional details are shown in Fig. 10.3.11.

Where the limit for ramps is exceeded a backdrop manhole construction can be considered. This consists of a vertical 'drop' pipe with access for both horizontal and vertical rodding. If the pipework is of clay or concrete the vertical pipe should be positioned as close to the outside face of the manhole as possible and encased in not less than 150 mm of mass concrete. Cast iron pipework is usually sited inside the chamber and fixed to the walls with holderbolts. Whichever material is used the basic principles are constant (see Fig. 10.3.11).

Changing the gradient of the incoming drain to bring its invert level in line with that of the sewer requires careful consideration and design. Although simple in conception the gradient must be such that a self-cleansing velocity is maintained and the requirements of Building Regulation Part H are not contravened.

Connections of small diameter drains to large diameter sewers can be made by any of the methods described above or by using a saddle connection. A saddle is a short socketed pipe with a flange or saddle curved to suit the outer profile of the sewer pipe. To make the connection a hole must be cut in the upper part of the sewer to receive the saddle ensuring that little or no débris is allowed to fall into the sewer. A small pilot hole is usually cut first and this is enlarged to the required diameter by careful cutting and removing the débris outwards. The saddle connection is bedded on to the sewer pipe with a cement mortar and the whole connection surrounded with a minimum of 150 mm of mass concrete (see Fig. 10.3.12).

DRAIN TESTING

To satisfy the Building Regulation requirements for watertightness of drains, reference to Approved Document H1 permits application of either a water or an air test. A smoke test could also be applied to determine the position of any apparent

Brick-levelling course

600 × 600 cast iron cover
and frame to BS EN 124

Ramp formed in
concrete benching

100 mm precast
concrete slab

Flow

Wall in BS 3921
class B bricks
in cm. mt. (1:3)
laid to English
Bond

700 max.

Mass concrete
benching

Incoming
drain

Mass concrete base

Private or public sewer

Ramp connection

150 × 150 cast iron
cover and frame

60 × 600 cast iron cover
and frame to BS EN 124

Incoming
drain

Flow

100 mm precast
concrete slab

Double
square
junction

Galv. step irons
to BS 1247 every
4th course vertically
set staggered at
230 c/c horizontally
1st step 450 mm
below cover level

Lever
locking
stopper

Drop
pipe

Wall in BS 3921
class B bricks
in cm. mt. (1:3)
laid to English
Bond

150 mm min.
concrete
encasing

90° bend

Mass concrete
benching

Mass concrete base

Private or public sewer

Backdrop connection

Figure 10.3.11   Manhole and inspection chamber sewer connections

150 mm concrete encasing
to saddle connection

Shoulder of saddle
shaped to fit outside
diameter of sewer and
to form oblique
junction

Flow

Incoming
drain

Flow

Private or public sewer

Connection arrangement

Incoming drain

Rigid joint

Flow

Saddle bedded
din cm. mt. (1:3)

At least
half saddle
width to
bear on
sewer wall

45° 45°

Private or
public sewer

Note: Saddle connection should be made in
the crown of the sewer pipe within 45°
on either side of the vertical axis

Typical section

Figure 10.3.12   Saddle connections to sewers

leakage revealed by the other tests. The Approved Document also makes reference
to the recommendations contained in BS 8301 which is the Code of Practice for
building drainage.

   The local authority will carry out drain testing after the backfilling of the drain
trench has taken place, therefore it is in the contractor's interest to test drains and

private sewers before the backfilling is carried out since the detection and repair of any failure discovered after backfilling can be time-consuming and costly.

TYPES OF TEST

There are three methods available for the testing of drains:

- ■ **Water test**   The usual method employed and is carried out by filling the drain run being tested with water under pressure and observing if there is any escape of water (see Fig. 10.3.13).
- ■ **Smoke test**   This is a dated practice, but if required smoke can be pumped into the sealed drain run under test and any fall in pressure as indicated by the fall of the float on the smoke machine observed (see Fig. 10.3.14).
- ■ **Air test**   This is not a particularly conclusive test but it is sometimes used in special circumstances such as large diameter pipes where a large quantity of water would be required. If a failure is indicated by an air test the drain should be retested using the more reliable water test (see Fig. 10.3.15).

The illustrations of drain testing have been prepared on the assumption that the test is being carried out by the contractor before backfilling has taken place.

In general the testing of drains should be carried out between manholes; manholes should be tested separately and short branches of less than 6.000 m should be tested with the main drain to which they are connected; long branches would be tested in the same manner as a main drain.

SOAKAWAYS

A soakaway is a pit dug in permeable ground which receives the rainwater discharge from the roof and paved areas of a small building and is so constructed that the water collected can percolate into the surrounding subsoil. To function correctly and efficiently a soakaway must be designed taking into account the following factors:

1. Permeability or rate of dispersion of the subsoil.
2. Area to be drained.
3. Storage capacity required to accept sudden inflow of water such as that encountered during a storm.
4. Local authority requirements as to method of construction and siting in relation to buildings.
5. Depth of water table.

Before any soakaway is designed or constructed the local authority should be contacted to obtain permission and ascertain its specific requirements. Some authorities will only permit the use of soakaways as an outfall to a subsoil drainage scheme.

Expanding ring

Cap

Wing nut

Plates

Flat moulded rubber expanding ring

13 or 25 mm dia. brass threaded outlet

Brass cap

Brass wing nut

Galvanized iron plates

Pipe and plug

Typical drain plug details

Temporarily jointed vertical length(s) of pipe to give required test head of 1.5 m which should be held for 2 hours after initial drop in water level and subsequent topping up – measure leakage over a 30 minute period

Drain plug to lower end of run under test

Drain under test filled with water

Bedding

Temporarily jointed bend

Figure 10.3.13   Water testing of drains

Removable float guide

Copper cover or float

Square copper tank filled to
within 6 mm of top with water

Steel tube
outlet

Valve

Bellows handle
with security
chain

Leather covered
double action
bellows

502 × 228 × 223
overall frame

32 mm bore
embedded
wire rubber
tube or
25 mm bore
flexible
metallic
tube

Smoke testing
machine

Expanding
plug type
connection

Drain under test filled
with smoke

Bedding

Drain plug capped or stopper
inflated when smoke has reached
upper end of drain run

Smoke machine details

Smoke is produced in the fire box by
burning oily cotton waste or prepared
smoke paper – after a few strokes
of the bellows the float rises and
will remain stationary if drain run is sound

Figure 10.3.14 Smoke testing of drains

Tube connection nipple

Tube leg connected to hand bellows and rubber tube

Door

Water in 'u' tube showing 75 mm

Cap

Tube leg open to atmosphere

Metal box

Manometer or 'u' gauge

Manometer

Hand bellows and rubber tube

Expanding plug type connection

Drain under test

Bedding

Diameters up to 300 mm available

Canvas covered rubber air bag

Hand pump

1.400 long rubber tube

Air bag stopper

After pumping has ceased the test pressure of 100 mm shown on the manometer should not fall below 75 mm in 5 minutes

Drain plug or air bag stopper

Figure 10.3.15 Air testing of drains

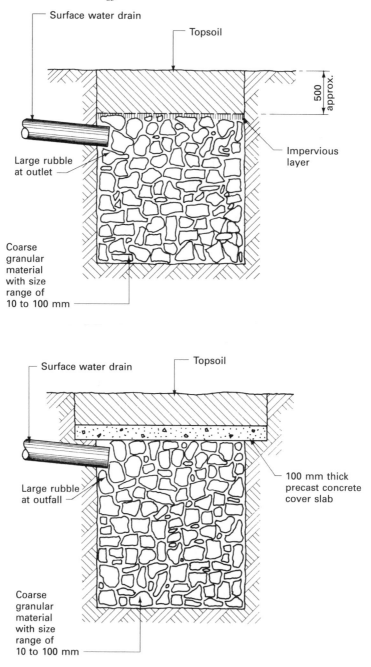

Surface water drain

Topsoil

500 approx.

Impervious layer

Large rubble at outlet

Coarse granular material with size range of 10 to 100 mm

Surface water drain

Topsoil

100 mm thick precast concrete cover slab

Large rubble at outfall

Coarse granular material with size range of 10 to 100 mm

Suitable coarse granular materials include broken bricks, crushed sound rock and hard clinker

Figure 10.3.16   Filled soakaways

The rate at which water will percolate into the ground depends mainly on the permeability of the soil. Generally clay soils are unacceptable for soakaway construction, whereas sands and gravels are usually satisfactory. An indication of the permeability of a soil can be ascertained by observing the rate of percolation. A bore hole 150 mm in diameter should be drilled to a depth of 1.000 m. Water to a depth of 300 mm is poured into the hole and the time taken for the water to disperse is noted. Several tests should be made to obtain an average figure and the whole procedure repeated at 1.000 m stages until the proposed depth of the soakaway has been reached. A suitable diameter and effective depth for the soakaway can be obtained from design guidance in Building Research Establishment Digest 365.

An alternative method, using a simple empirical formula, calculates the volume of a soakaway by allowing for a storage capacity equal to one-third of the hourly rain falling onto the area to be drained; that is,

$$C = \frac{AR}{3}$$

where C = capacity (m^3)
   A = area to be drained (m^2)
   R = rainfall (m/h)

For example:

Area to be drained = 150 m^2
storm water allowance = 75 mm/h

therefore

$$C = \frac{150 \times 0.075}{3} = 3.75\,\text{m}^3$$

TYPES OF SOAKAWAY

A soakaway is constructed by excavating a pit of the appropriate size and either filling the void with selected coarse granular material or alternatively lining the sides of the excavation with brickwork or precast concrete rings (see Figs 10.3.16 and 10.3.17).

Filled soakaways are usually employed only for small capacities, since it is difficult to estimate the storage capacity and the life of the soakaway may be limited by the silting up of the voids between the filling material. Lined soakaways are generally more efficient, have a longer life and if access is provided can be inspected and maintained at regular intervals.

Soakaways should be sited away from buildings so that foundations are unaffected by the percolation of water from the soakaway. The minimum 'safe' distance is often quoted as 5.000 m but local authority advice should always be sought. The number of soakaways required can only be determined by having the facts concerning total drain runs, areas to be drained and the rate of percolation for any particular site.

Brick-levelling courses

600 × 600 cast iron cover and frame

Surface water drain

100 mm thick precast concrete cover slab

Brick walls laid in honeycomb bond

Alternatives for walls – dry rubble walling and dry jointed brickwork

Mass *in-situ* concrete strip

Brick-lined soakaway

Brick-levelling courses

Cast iron cover and frame

Drain

Precast concrete light duty chamber cover slab

600 min. diameter

Dry ogee joints

675 to 1.800 dia.

300 to 1.200 in multiples of 150 mm

Granular backfill

75 mm dia. drainage holes

Precast concrete cylindrical chamber sections

Mass *in-situ* concrete strip

Precast concrete soakaway

Figure 10.3.17   Lined soakaways

DOMESTIC ELECTRICAL INSTALLATIONS

A simple definition of the term *electricity* is not possible but it can be considered as a form of energy due to the free movement of tiny particles called *electrons*. If sufficient of these free or loose electrons move an electric current is produced in the material in which they are moving. Materials such as most metals and water which allow an electric current to flow readily are called *conductors* and are said to have a low resistance. Materials which resist the flow of an electric current such as rubber, glass and most plastics are called *insulators*.

For an electric current to flow there must be a complete path or circuit from the source of energy through a conductor back to the source. Any interruption of the path will stop the flow of electricity. The pressure which forces or pushes the current around the circuit is called the *voltage*. The rate at which the current flows is measured in *amperes* and the resistance offered by the circuit to passage of electrons is measured in *ohms*. A *watt* is the unit of power and is equal to the product of volts × amperes; similarly it can be shown that voltage is equal to the product of amperes × ohms.

Another effect of an electric current flowing through a conductor is that it will dissipate wasted energy in the form of heat according to the resistance of the conductor. If a wire of the correct resistance is chosen it will become very hot and this heating effect can be used in appliances such as cookers, irons and fires. The conductor in a filament bulb is a very thin wire of high resistance and becomes white hot thus giving out light as well as heat.

Most domestic premises receive a single-phase supply of electricity from an area electricity authority at a rating of 230 volts, and a frequency of 50 hertz. The area authority's cable, from which the domestic supply is taken, consists of four lines, three lines each carrying a 230 volt supply and the fourth is the common return line or neutral which is connected to earth at the transformer or substation as a safety precaution should a fault occur on the electrical appliance. Each line or phase is tapped in turn together with the neutral to provide the single-phase 230 V supply. Electricity is generated and supplied as an *alternating current*, which means that the current flows first one way then the other; the direction change is so rapid that it is hardly discernible in such fittings as lights. The cycle of this reversal of flow is termed *frequency*.

External cavity wall

Consumer unit (fuse box)

Plastic meter cupboard

Phase and neutral intake to sealing chamber (see Fig. 10.4.4)

Ground floor

Plastic protective conduit

450 min.

Figure 10.4.1   Underground domestic electrical supply

Figure 10.4.1 shows a typical underground domestic electricity supply to an external meter box. The consumer unit should be located as close as possible to the intake on the inside of the cavity wall as shown, or on an adjacent partition.

The conductors used in domestic installations are called cables and consist of a conductor of low resistance such as copper or aluminium surrounded by an insulator of high resistance such as rubber or plastic. Cable sizes are known by the nominal cross-sectional area of the conductor and up to 2.5 mm^2 are usually of one strand. Larger cables consist of a number of strands to give them flexibility. All

Ground floor plan

Upper floor plan

Symbols		
✓ One-way switch	⚡ Pendant switch	☑ Consumer unit
✗ Two-way switch	◯ Ceiling outlet	⦿ Meter
▷— Switch socket outlet	⊶ Immersion heater	© Cooker control

Figure 10.4.2 Typical domestic electrical layout

cables are assigned a rating in amperes which is the maximum load the cable can carry without becoming overheated.

For domestic work, wiring drawings are not usually required; instead the positions of outlets, switches and lighting points are shown by symbols on the plans. Specification of fittings, fixing heights and cables is given either in a schedule or in a written document (see Fig. 10.4.2).

CIRCUITS

Domestic buildings are wired using a ring circuit as opposed to the older method of having a separate fused subcircuit to each socket outlet. Lighting circuits are carried out by using the 'loop in' method.

The supply or intake cable will enter the building through ducts and be terminated in the area authority's fused sealing chamber which should be sited in a dry accessible position. From the sealing chamber the supply passes through the meter, which records the electricity consumed in units of kilowatt/hours, to the consumer unit which has a switch controlling the supply to the circuit fuses or miniature circuit breakers. These fuses or circuit breakers are a protection against excess current or overload of the circuit since, should overloading occur, the fuse or circuit breaker will isolate the circuit from the supply.

The number of fuseways or miniature circuit breakers contained in the consumer unit will depend upon the size of the building and the equipment to be installed. A separate ring circuit of 32 amp loading should be allowed for every 100 m^2 of floor area and as far as practicable the number of outlets should be evenly distributed over the circuits. A typical domestic installation would have the following circuits from the consumer unit:

1. 6 amp: ground floor lighting up to ten fittings or a total load of 6 amps.
2. 6 amp: upper floor lighting as above.
3. 16 amp: immersion heater.
4. 32 amp: ring circuit 1.
5. 32 amp: ring circuit 2.
6. 32 amp: ring circuit, kitchen.
7. 45 amp: cooker circuit.

[*Note*: A further 40 or 45 amp circuit breaker or fuse may be installed for an electric shower unit.]

The complete installation is earthed by connecting the metal consumer unit casing to the sheath of the supply cable or by connection to a separate earth electrode.

For lighting circuits using sheathed wiring a 1.5 mm^2 conductor is required and therefore a twin with earth cable is used. The 'loop-in' method of wiring is shown diagrammatically in Fig. 10.4.3. It is essential that lighting circuits are properly earthed since most domestic fittings and switches contain metal parts or fixings which could become live should a fault occur. Lighting circuits using a conduit installation with single-core cables can be looped from switch to switch as shown in Fig. 10.4.3. Conduit installation consists of metal or plastic tubing together with various boxes for forming junctions and housing switches which gives a protected rewireable system. If steel-screwed conduit is used it will also serve as the earth leakage path but if plastic conduit is used, being non-conductive, the circuit must have an insulated earth conductor throughout.

A ring circuit for socket outlets consists of a twin 2.5 mm^2 earthed cable starting from and returning to the consumer unit. The cables are looped into the outlet

'Loop-in' method of wiring using sheathed cable

Single-core cable looped from switch to switch

Figure 10.4.3   Typical lighting circuits

boxes making sure that the correct cable is connected to the correct terminals (see Fig. 10.4.4). The number of outlets is unlimited if the requirement of 1 ring circuit per 100 m² of floor area has been adopted. Spur outlets leading off the main ring circuit are permissible provided the limitation of not more than two outlet sockets on any one spur and not more than half the socket outlets on the circuit are on spurs is not exceeded. Socket outlets can be switched controlled and of single or double outlet; the double outlet is considered the best arrangement since it

Main switch

6 amp mcb – ground floor lighting circuit
6 amp mcb – upper floor lighting circuit
16 amp mcb – immersion heater
32 amp mcb – ring circuit, ground floor
32 amp mcb – ring circuit, upper floor
32 amp mcb – ring circuit, kitchen
45 amp mcb – cooker

Meter

Sealing chamber

Supply cable

Consumer unit

Socket outlets

Line

Neutral

Earth

Earth terminal – cable colour green/yellow

Socket outlet

Spur outlet

BS 1363 plug

Cap screw

Cartridge fuse – 3 amp for up to 720 watt rating
13 amp for up to 3,000 watt rating

Line terminal – cable colour brown

Cable grip

Neutral terminal – cable colour blue

Figure 10.4.4   Ring circuits and plug wiring

discourages the use of multiple adaptors. Fixed appliances such as wall heaters should be connected directly to a fused spur outlet to reduce the number of trailing leads. Moveable appliances such as irons, radios and standard lamps should have a fused plug for connection to the switched outlet, conforming to the requirements of BS 1363. The rating of the cartridge fuse should be in accordance with rating of the appliance. Appliances with a rating of not more than 720 watts should be protected by a 3 amp fuse and appliances over this rating up to 3,000 watts should have a 13 amp fuse. As with the circuit, correct wiring of the plug is essential (see Fig. 10.4.4).

The number of outlets is not mandatory but the minimum numbers recommended for various types of accommodation are:

- Kitchens: 6 plus cooker control unit with one outlet socket.
- Living rooms: 8.
- Dining rooms: 4.
- Bedrooms: 4.
- Halls: 1.
- Landings: 1.
- Garages: 1.
- Stores and workshops: 1.

The outlets should be installed around the perimeter of the rooms in the most convenient and economic positions to give maximum coverage with minimum amount of trailing leads.

Cables sheathed with PVC can be run under suspended floors by drilling small holes on the neutral axis of the joists; where the cables are to be covered by wall finishes or floor screed they should be protected by either oval conduit or by means of small metal cover channels fixed to the wall or floor. Systems using mineral-insulated covered cables follow the same principles. This form of cable consists of single strands of copper or aluminium all encased in a sheath of the same metal which is densely packed with fine magnesium oxide insulation around the strands. This insulating material is unaffected by heat or age and is therefore very durable but it can absorb moisture. The sealing of the ends of this type of cable with special sealing 'pots' is therefore of paramount importance. Cables used in a conduit installation have adequate protection but it is generally necessary to chase the walls of the building to accommodate the conduit, outlet socket boxes and switch boxes below the wall finish level. Surface run conduit is normally secured to the backing by using screwed shaped clips called *saddles*.

Guidance for the installation of electrical circuits and electrical equipment is provided in the Building Regulations, Approved Document P: Electrical safety. Further details are produced in the Institution of Electrical Engineers Regulations in accordance with BS 7671: Requirements for electrical installations.

DOMESTIC GAS INSTALLATIONS

Gas is a combustible fuel which burns with a luminous flame; it is used mainly in domestic installations as a source of heat in appliances such as room heaters, cookers and water heaters. Gas can also be utilized to provide the power for washing machines and refrigerators; the use of gas as a means of artificial lighting has been superseded by electricity.

Gas is supplied by area authorities under the general co-ordination and guidance of the Gas Council. The supply was originally a manufactured gas known as 'town gas' but this has been replaced by off-shore supplies of natural gas, often referred to as 'North Sea Gas'. Town gas can be processed from coal, oil or imported natural gas and has a high hydrogen content of approximately 50% with a calorific value of about 18.6 MJ/m^3. Natural gas has no hydrogen content but has a very high percentage content of methane of approximately 95% with a calorific value of about 37.3 MJ/m^3. The pressure burning rate and amount of air required for correct combustion varies with the two forms of gas supply and therefore it is essential that the correct type of burner is fitted to the appliance.

The installation of a gas service to a domestic building is usually carried out by the local area authority's contractors and can be considered in three distinct stages:

Main This is part of the gas distribution system, which is interconnected to form a grid. The grid arrangement enables maintenance to be undertaken in isolation without disturbing the whole area. It is laid and maintained by the local gas authority or a subsidiary pipeline company. For identification purposes gas pipelines are colour coded over their length or in 150 mm wide bands with yellow ochre.

Service pipe The connection pipe between the main and the consumer control which is positioned just before the governor and meter. In small domestic installations the service pipe diameter is 25–50 mm according to the number and type of appliances being installed. The pipeline should be as short as possible, at right angles to the main, laid at least 375 mm below the ground to avoid damage. Figure 10.5.1 shows a typical domestic installation with meter details and accessories.

Figure 10.5.1 Domestic gas supply

Internal installation   This commences at the consumer control valve and consists of a governor to stabilize the pressure and volume, the meter which records the volume of gas consumed, pipework to convey the gas supply to the appliances.

PIPEWORK

Pipework can be of mild steel, solid drawn copper pipes and flexible tubing of metal reinforced rubber for use with portable appliances such as gas pokers. The size of installation pipes will depend upon such factors as gas consumption of appliances, frictional losses due to length of pipe runs and bends. Gas pipes are fixed by means of pipe hooks, clips or ring brackets at approximately 1.500 m centres. All exposed steel pipes should be protected with bituminous paint. Pipes which pass through walls are housed in a sleeve of non-corrodible material surrounded by packing of incombustible material such as glass fibre to facilitate easy replacement and to accommodate small differential movements.

APPLIANCES

Gas appliances fall mainly into two groups:

- Gas supply only: refrigerators, radiant heaters, gas pokers.
- Gas supply plus other services: central heating units, water heaters.

Gas refrigerators are silent, reliable, requiring little or no maintenance and cheap to run. They work on the absorption principle using ammonia dissolved in water. Free-standing, built-in and table models are available; all require a firm level surface, a well-ventilated position, clearance for opening of doors and access to the rear to make the fixed connection to the supply. Gas room heaters can be radiant heaters where the heat source is visible, convector heaters or in combination as a radiant-convector heater. Room heaters using gas have high efficiency, rapid heat build-up, are easy to clean and have low maintenance. Gas fires or room heaters will require a chimney and/or a flue complying with the requirements of Building Regulations Part J. Approved Document J provides guidance on acceptable installation practice and procedures with details of flue sizes, materials and location. Gas cookers are an exception to the requirements for flues, but they should only be installed and used in well-ventilated kitchens. Room-sealed or balanced flue heaters are appliances which have the heater body sealed from the room; the heater obtains air for combustion by an inlet connected directly with the external air and return the products of combustion to the external air by a separate outlet but usually to a common grill or guard. The termination of the flues are so designed that the external wind pressure effects are balanced, which obviates the need for a traditional flue (see Chapter 5.2).

Gas central heating consists of a boiler, connected to a flue complying with the Building Regulations, which provides the heating source for a fanned warm air system which is flexible in design and has a quick response or a circulated hot water system.

The installation of gas services, like those of electricity, are specialist subjects of building technology and normally form the basis of separate study. For a greater appreciation and development of these topics see *Building Services, Technology and Design* by R. Greeno (Longman and CIOB, co-publishers).

BIBLIOGRAPHY

Barry, R. *The Construction of Buildings*. Crosby Lockwood & Sons Ltd.

Elder, A. J. *A. J. Guide to the Building Regulations*. The Architectural Press.

Elder, A. J. *Relevant A. J. Handbooks*. The Architectural Press.

Fisher Cassie, W. and Napper, J. H. *Structure in Building*. The Architectural Press.

Greeno, R. *Building Services, Technology and Design*. Addison Wesley Longman and CIOB.

Greeno, R. *Principles of Construction*. Addison Wesley Longman.

Hall, F. *Plumbing*. Macmillan.

Handisyde, Cecil C. *Building Materials*. The Architectural Press.

Leech, L. V. *Structural Steelwork for Students*. Butterworth.

Llewelyn Davies, R. and Petty, D. J. *Building Elements*. The Architectural Press.

McKay, W. B. *Building Construction*, Vols 1 to 4. Longman.

Ragsdale, L. A. and Raynham E. A. *Building Materials Technology*. Edward Arnold Ltd.

Wooley, L. *Drainage Details*. Chapman & Hall.

GENERAL

Application of Mastic Asphalt. Mastic Asphalt Council and Employers Federation.

Building Regulations, England and Wales. Stationery Office.

Copper Roofing. Copper Development Association.

Gas Handbook for Architects and Builders. The Gas Council.

Handbook on Structural Steelwork. The British Constructional Steelwork Association Ltd and The Steel Construction Institute.

Mitchells Building Series. Addison Wesley Longman.

'Sanitation details by Aquarius', *Building Trades Journal*. Chapman & Hall.

The Blue Book of Plasterboard. British Gypsum Ltd.

The Green Book of Plasterboard. British Gypsum Ltd.

The Plumber's Handbook. Lead Development Association.

526

OTHER READING

Relevant advisory leaflets. DoE.
Relevant BRE Digests. Construction Research Communications Ltd.
Relevant British Standards Codes of Practice. British Standards Institution.
Relevant British Standards. British Standards Institution.
Relevant manufacturers' catalogues contained in the Barbour Index and Building
 Products Index Libraries.

INDEX